Routledge Revivals

Alternative Arrangements
for Marine Fisheries

The maintenance of the freedom of fishing in the face of the changing circumstances that were occurring at the time of this title's original publication in 1973 had produced several damaging consequences. It had led to considerable waste, in both biological and economic terms, and had been the source of increasing conflict. This waste can only be prevented by the adoption of management measures and by the removal of the condition of free and open access. This book explores various techniques for this, and will be of interest to students of environmental management.

Alternative Arrangements for Marine Fisheries

An Overview

Francis T. Christy, Jr.

First published in 1973
by Resources for the Future, Inc.

This edition first published in 2016 by Routledge
2 Park Square, Milton Park, Abingdon, Oxon, OX14 4RN

and by Routledge
711 Third Avenue, New York, NY 10017

Routledge is an imprint of the Taylor & Francis Group, an informa business

© 1973 Resources for the Future, Inc.

Publisher's Note
The publisher has gone to great lengths to ensure the quality of this reprint but points out that some imperfections in the original copies may be apparent.

Disclaimer
The publisher has made every effort to trace copyright holders and welcomes correspondence from those they have been unable to contact.

A Library of Congress record exists under LC control number: 73020841

ISBN 13: 978-1-138-94118-2 (hbk)
ISBN 13: 978-1-315-67327-1 (ebk)
ISBN 13: 978-1-138-94216-5 (pbk)

ALTERNATIVE ARRANGEMENTS
FOR MARINE FISHERIES
AN OVERVIEW

Francis T. Christy, Jr.

RESOURCES FOR THE FUTURE, INC.
Washington, D.C.

May 1973

Contents

Figures

Introduction

FOR THE PAST THREE HUNDRED YEARS, one of the basic principles of the freedom of the seas has been the freedom of fishing. Under this principle, states have generally maintained relatively narrow limits of jurisdiction and fishermen have had free and open access to all stocks on the high seas beyond the national limits. In these wide international waters, no individual state or group of states has had a right to exclude others from freely enjoying the common resources.

In the next few years, however, the principle of the freedom of fishing will disappear over most areas of the oceans. This process of change, initiated after the Second World War, is both inevitable and desirable. It may, however, be accompanied by severe conflict and by a waste of the benefits from ocean fisheries. As states meet at the third UN Conference on the Law of the Sea for the purpose of establishing new fisheries regimes, their chief task will be to facilitate the change that is occurring in such a way that both conflict and waste are minimized.

In the early 1600's, Hugo Grotius advanced two major arguments to support the freedom of fishing. He maintained that fishery resources were so abundant that there was no value in having exclusive rights; that all could use the resources without diminishing the returns to any one user. Such conditions—as, for example, with the air we breathe—mean that exclusive rights are valueless since no one is willing to pay for something that can be obtained elsewhere at no cost. As a second reason, Grotius maintained that the acquisition of exclusive rights was infeasible because of extensive migratory movements of the fish and the inability of coastal states to exercise jurisdiction over large areas of the sea. In essence, these arguments asserted that the benefits of acquiring jurisdiction were not worth the costs.

These two arguments have been staunchly supported by the many descendents of Grotius. And even today, there are those who argue that the freedom of the seas for fishing is the best of all regimes and produces the greatest benefits for the world community. They frequently do so, however, on the basis of questionable motives and dubious reasons. The myth of abundance of the sea's wealth in fisheries can no longer be maintained in the face of the growing record of depletion of stocks and the increasing conflicts over the scarce resources. The difficulties of acquiring satisfactory rights of jurisdiction persist, and will continue to plague arrangements for the management of fisheries. However, the difficulties have not precluded attempts to acquire jurisdiction, as is evident in the enlargement of coastal state claims and in the growth in multilateral arrangements that serve to exclude newcomers. Nor do they mean that the acquisition and exercise of a satisfactory degree of authority is impossible.

The maintenance of the freedom of fishing in the face of the changed circumstances has produced several damaging consequences. It has led to considerable waste, in both biological and economic terms. And it has been the source of increasing conflict. The waste, discussed more fully in Chapter I, is due to the fact that under the condition of free and open access, no individual user is able to restrain his catch in the interests of future returns, because anything he leaves in the sea for tomorrow will be taken by others today.

The waste can only be prevented by the adoption of management measures and by the removal of the condition of free and open access. The various techniques for this, discussed in Chapter II, have quite different implications for the benefits sought from fisheries and for the costs involved in making them effective. However, it is clear from that discussion that a much greater degree of authority and control must be exercised than has been available in the past under the principle of the freedom of fishing.

But the acquisition and exercise of such authority requires states to address, directly and definitively, the problems of wealth distribution. These problems—the subject of Chapter III—can no longer be avoided under the pretense that all nations of the world benefit from unrestricted access to a common resource. But even were it not for the needs of management, the problems of distribution could no longer be ignored because the values of acquiring exclusive rights are now great enough to make states willing to incur the costs and difficulties. Such difficulties can be very great, particularly in the frequent situations where stocks of fish swim freely across man-made boundaries and must be jointly managed if they are to be managed at all. In Chapter IV, there is some discussion of the ways in which states might organize themselves in creating institutions to deal with the functions of research, regulation, enforcement, and the settlement of disputes.

While the problems of developing and maintaining satisfactory arrangements for complex fisheries may appear to be overwhelming, not all of them need to be resolved at the same time, nor are all of them equally relevant to all regions. Many of the problems raised in this paper refer to specific situations, and will need to be addressed only at local or regional levels. Others, however, are of a more general nature, and are discussed in the final chapter as items that might be raised at the UN Conference on the Law of the Sea. An attempt has been made to identify some of the more important general principles that might be proposed and to examine them in terms of their value in improving the production of benefits from fisheries and in reducing the potential for conflict.

I Fundamental Characteristics of Fisheries

THERE ARE CERTAIN FUNDAMENTAL CHARACTERISTICS of fisheries that are of critical importance for decisions on both management and distribution. These characteristics distinguish fisheries from most other forms of enterprise and they create particular difficulties for the establishment of effective and appropriate arrangements. The characteristics include (a) the migratory behavior of fish; (b) the fact that they are wild stocks not readily subject to cultivation and that the yield of any particular stock is limited; (c) the complex interrelationships between fish species and fishing effort; and (d) the general treatment of fish as common property natural resources for which access is both free and open. In addition, because of these characteristics and the history of the freedom of the seas, there tends to be confusion between the objectives of management and those of wealth distribution—a confusion that often obscures the range of alternative solutions.

Before turning to a discussion of the characteristics of fisheries, it may be useful to define some of the terms used in this paper. The term "fisheries" should be considered as a generic term relating to all kinds of living marine resources utilized by man. As thus used, it includes marine mammals such as whales and seals, molluscs, crustaceans, and other animals of the ocean environment, even though these are not, strictly speaking, fish.

A distinction is made between the terms "species" and "stocks." The former refers to a particular kind of living marine resource wherever it occurs. The latter refers to a particular group of fish of a certain species—a group that is sufficiently isolated from other groups of the same species to permit separate management practices. For example, a particular species of tuna may be found in several places throughout the world's oceans, while an identifiable stock of that species will be found only in a certain location. The distinction is important because any particular stock might require quite different management practices than other stocks of the same species.

The term "demersal" refers to species that feed on the bottom and that are thus closely associated with the relatively shallow waters of the continental shelves and slopes. "Pelagic" fish are those that feed on the surface. Some may occur in shallow waters while others may be found where the waters are very deep.

MIGRATORY BEHAVIOR

There is considerable variation in the migratory patterns of living marine organisms—ranging from sedentary species (such as oysters), which remain fixed in place during most of their life cycles, to highly migratory species (such as certain tunas) which may swim across entire oceans. Some species migrate in the relatively shallow waters over the continental shelf, moving freely between the fishery zones claimed by neighboring coastal states. Others move from the waters of a single coastal state out onto the high seas, many hundreds of miles away from land. And some migrate vertically from the deep waters of the continental slope to the shallow waters of the shelf.

The fact that many valued species of fish move freely through national and international waters presents one of the major sources of difficulties for arrangements governing the management and allocation of fisheries wealth. It leads to the primary and fundamental necessity for unification of control among the states in whose waters the resources can be found. Without such unification, there is no incentive for any individual state to engage in conservation practices by itself, because to restrain its own catch would be to leave greater catches for other, nonconserving states. Thus, where stocks are shared, agreement among the states is absolutely necessary if waste is to be prevented.

Classification of Migratory Characteristics

The migratory characteristics of different species can be classified into four divisions, although the classification is not precise and, as noted below, may be accompanied by fairly severe problems. The four divisions adopted here are (a) sedentary, (b) anadromous, (c) highly migratory oceanic, and (d) coastal.

Sedentary species were defined and given special treatment in the 1958 Geneva Convention on the Continental Shelf. They were defined as those marine organisms which "at the harvestable stage, either are immobile on or under the seabed or are unable to move except in constant physical contact with the seabed or subsoil." This clearly includes oysters, clams, and other kinds of molluscs. Certain crusta-

2

ceans, such as crabs, lobsters, and shrimp, may or may not be included in this definition. According to the 1958 Convention, sedentary species are considered as natural resources of the continental shelf and fall within the exclusive jurisdiction of the adjacent coastal state in the same manner as the minerals of the shelf bed.

Anadromous species are those organisms that spawn in fresh water but spend a significant part of their life cycle in the oceans. In some cases, such as salmon, they may swim thousands of miles in the high seas during their oceanic phase. A related group of marine organisms are those that are catadromous—spawning on the high seas but spending much of their lives in fresh water. Certain eels are found in this category. In both cases, the species might be considered as highly migratory oceanic species during part of their life cycle, while during another part they are clearly subject to control, management, and capture by individual coastal states.

Highly migratory, oceanic species have two important characteristics: they swim great distances and they frequently are found far from land. These characteristics are not precisely defined, and the distinction between highly migratory oceanic species and coastal species is far from clear. However, it is generally assumed that this category includes certain species of tuna, whales, and swordfish and other billfish.

The fourth category is that of coastal species. As noted above, there is no precise definition of this class. The general assumption is that it generally includes the species found relatively close to land during most of their life cycles. They may or may not migrate over great distances. The great majority of marine organisms harvested by man fall into this category.

Migratory Behavior and the "Species Approach"

It may be apparent that there is a close relationship between the classification described above and the "species approach" to the resolution of fisheries problems as proposed by the United States and others. It is suggested in the U.S. proposal that three of the four categories be given separate treatment in the new arrangements and regimes for fisheries, with the fourth (sedentary species) already receiving separate treatment through the 1958 Convention. While there are several important reasons for adopting this suggestion, there are severe problems of definition that need to be worked out and there are major difficulties involved in determining the kind and nature of the separate treatment that might be granted. Some of these are discussed below for each of the four categories.

3

With regard to sedentary species, many controversies have already arisen over definition. Thus far, the controversies have related to the determination of whether or not (or the degree to which) the animal is "unable to move except in constant physical contact with the seabed or subsoil." There are differences of opinion, for example, over king crab off Alaska and lobster off Brazil. It would appear that these controversies cannot be readily resolved unless the definition is made more precise or unless agreement is reached on a list of species generally accepted as being sedentary.

An additional problem with the definition of sedentary species lies in the absence of a clear limit of the seaward extent of a coastal state's jurisdiction. Certain species that might be included in the definition are already being harvested in depths much deeper than 200 meters. Shrimps and prawns are being commercially taken in depths of 700 meters in the Mediterranean, 600 meters off the coast of Angola and in the North Atlantic, and down to 500 meters off the coast of Norway. Lobsters have been found at depths of 900 meters and crabs down to 730 meters, although production at these depths does not appear to be presently commercial.[1]

There is apparently little present disagreement over the kind of special treatment to be accorded to sedentary species. As noted, the 1958 Convention on the Continental Shelf provides the adjacent coastal state with fully exclusive rights over such resources, identical to the rights over seabed minerals. But the question might be raised as to whether the same degree of rights is acceptable for the sedentary species taken from the deeper waters of the continental slope. With regard to the deep water minerals of the continental slope, the U.S. "Draft United Nations Convention on the International Seabed Area" provides for a sharing of revenues between the coastal state and the world community. It specifies, however, that no such limits on jurisdiction will be accorded to the "living resources of the seabed." However, if the principle of sharing of revenues is adopted for miner-

[1]Milner B. Schaefer, "Some Considerations of Living Resources Associated with the Deep Sea-Bed," in J. Sztucki, ed., *Symposium on the International Regime of the Sea-Bed* (Rome: Accadémia Nazionale dei Lincei, 1970). It might be noted that the USSR in 1968, undertook successful commercial trawling at depths ranging from 800 to 1,370 meters off the northern Newfoundland bank. This, however, was for grenadier, a demersal finfish, and not a sedentary species. See L. N. Pechenik and F. M. Troyanovskii, *Trawling Resources on the North Atlantic Continental Slope* (Translated from Russian by Israel Program for Scientific Translations, Jerusalem 1971, National Technical Information Service, U.S. Dept. of Commerce, Springfield, Virginia), p. 48.

als, some states may raise the question as to whether or not it should also apply to sedentary organisms.

The rationale for granting separate treatment for sedentary species would appear to be simply that such species, because of their restricted movement, are readily subject to controls exercised by the coastal state. They are more like minerals than fish in this respect and capable of being fully contained within man-made boundaries. If this is the only rationale for separate treatment, however, then a number of species of finfish might also fall into the same category, as well as many crustaceans with severely limited migratory patterns. This question of rationale needs to be examined in much greater detail if there is to be a workable definition of sedentary species or an acceptable list of species to be granted the kind of separate treatment described above.

The definition of the anadromous and catadromous category is also subject to disagreement and should be made more precise. At present, the most important species in this category are various kinds of salmon, which clearly meet the conditions of the definition. But it can be anticipated that there will be controversies in the future over species that may spawn in brackish waters or that may use bays and estuaries during parts of their life cycles but are found at relatively great distances from shore during other parts. Certain kinds of herring and shrimp, for example, might be considered as falling into this category. Here again, there is need for much greater precision in definition or for acceptance of a list of species that might be considered as requiring the kind of separate treatment that is suggested for salmon.

The rationale for separate treatment of such species is fairly clear. During a critical part of their life cycles, these species are particularly and uniquely vulnerable or subject to the activities of a single coastal state. Improvements in yields are possible through investments in habitat controls, reduction of pollution, provision of fish ladders in dams, construction of artificial spawning beds, and even through selective breeding of better strains. Such investments are not likely to be made by the host state unless there is some assurance of receiving an adequate return. If great amounts of a stock are taken by other states when the stock swims beyond the waters of national jurisdiction, there is little incentive for the host state to incur investments or sacrifice other values in order to produce greater yields. Thus, some kind of separate treatment of these kinds of fish is necessary if yields are to be improved.

This fact indicates in part (but only in part) the kind of treatment that might be accorded to such species. It indicates that the host state should be assured sufficient benefits to warrant its investments in improvement of the yields. But there are no clear-cut answers to the questions of how benefits are to be allocated or who, aside from the host state, should share in the benefits. The rationale for separate treatment does not necessarily mean that the host state should receive the *entire* benefits from the resource.

Problems of definition are even greater for the categories of highly migratory oceanic species and coastal species. Indeed, it is difficult to imagine any set of definitions that will provide a clear distinction between the two categories. Perhaps, as in the case of other categories, it may become desirable to resolve definitional problems by getting acceptance of lists of individual species that should fall into the different classes.

The need to resolve definitional problems is important because there is a rationale for granting separate treatment to each of these categories. With regard to the highly migratory oceanic species, three reasons can be suggested. First, for some species of marine organisms, such as Antarctic whales, there is no coastal state that has either the interest or the authority to invoke and enforce conservation measures. These stocks occur sufficiently far from coastal states during most of their life cycles to be considered non-national in character, bearing some similarity in this respect to the manganese nodules that lie on the floor of the deep sea. No individual state can readily use the threat of extension of jurisdiction in order to enforce management controls. The user states are freed from such pressures and, as in the case of the Antarctic whales, may decide that their immediate interests in catch are greater than their long-run interests in conserved stocks. This freedom to waste might be considered sufficiently damaging to the interests of the world community to justify some kind of special treatment at the UN Conference on the Law of the Sea.

Second, the wide ranging movements of some species (such as certain tunas) may be so great that the authority of coastal states would be ineffective in the adoption and enforcement of management measures. For example, off the western coast of Latin America, jurisdictional limits even as great as 200 miles are not sufficient to fully contain stocks of skipjack and yellowfin tuna, among others. Modern tuna vessels are capable of harvesting these stocks beyond the limits of 200 miles and would undoubtedly do so to a greater extent than at present if coastal state controls were to become excessively restrictive. In these situations, management cannot be left solely in the hands of the coastal states.

Finally, and perhaps most important, the global mobility of tuna vessels, together with the tremendous recent increase in their total number and capacity, means that management measures must be international in scope, through close coordination between the regional

6

arrangements, through some kind of international body, or through other means. Many tuna vessels are so highly mobile that they can fish in any area of the world's oceans. They are also frequently so highly specialized that they cannot readily turn to other kinds of fishing. The product is of high unit value and almost entirely consumed in a few wealthy countries, tending to support an international industry that is relatively indifferent as to the sources of supply. Because of these various factors, restrictions on the catching of tuna in one region of the world leads to an almost immediate transfer of fishing effort to other regions and to great pressures on the limited stocks.

The global effects of controls that are imposed in any one region require a global approach to the resolution of management problems. This fact, however, provides only a rough indication of the kind of approach or separate treatment that would be desirable for the category of highly migratory oceanic species. It says nothing about the kind of measures that might be imposed, the degree of authority that might be granted to an international body, or the means for distributing benefits among the interested states. Since these matters are discussed in considerable detail in a companion Study Paper, they are not considered further here.[2]

The greatest amount of fish currently being taken fall into the category of coastal species. There is little that can be said by definition, other than that these are the species that occur relatively close to land. They may be demersal species, related to the seabed, and occur in the waters over the continental shelves and slopes. They may be pelagic species (feeding on the surface) and occur in relatively shallow waters or in deep waters close to shore where there are major upwelling currents. They may be highly migratory along coastlines or around the basins of small seas. They may eventually subsume the category of sedentary species in future regimes for the law of the sea. Whether or not separate stocks of the coastal species fall within the waters of a single coastal state depends both upon the migratory behavior of the stocks and upon the longitudinal and latitudinal dimensions of the state's boundaries. But in many, if not most, cases the stocks of coastal species are found in the waters of two or more coastal states.

The treatment of coastal species presents some of the most difficult problems for the law of the sea. These might be divided into two categories. One set of problems relates to the degree of authority that

[2]Saul Saila and Virgil Norton, *Alternative Arrangements for the Management of Tuna* (Washington: Resources for the Future, Program of International Studies of Fishery Arrangements, forthcoming.)

can be exercised by the coastal state and the degree to which this authority might be constrained by universal principles. For example, should there be a universal requirement that no state has a right to deplete a fishery resource even though that resource falls entirely within its waters? Should there be a universal principle that a coastal state is required to provide access to stocks that its fishermen are not fully utilizing? Should landlocked states be guaranteed access to or a share of the benefits from the fishery resources in the waters of the neighboring coastal states? How far out should coastal state authority extend? These, and many other problems like them, will have to be addressed at the forthcoming UN Conference on the Law of the Sea.

The other sets of problems, with regard to coastal species, involve the relationships between and among the states. These include the problems of achieving cooperation in the management of stocks that swim in the waters of two or more neighboring coastal states. And they include the problems of the allocation of fishery resources or of the benefits from the resources. While the UN Conference may be able to provide some general principles that will help in the resolution of these problems, they are becoming increasingly severe and will continue to do so well beyond the termination of the Conference. The extension of limits of jurisdiction does not, by any means, resolve the difficulties. It may provide coastal states with a greater degree of authority, but it does not obviate the necessity for adopting management measures and for making decisions on the distribution of benefits.

There are three sets of situations in which the problems of coastal species arise to a greater or lesser extent: (a) where a stock falls entirely within the waters of a single coastal state; (b) where a stock falls entirely within the waters of two or more coastal states; and (c) where a stock falls both within and outside of the waters of a state or group of states. In these last two situations, which prevail throughout many areas of the world, unification of management among the states sharing the resource is a fundamental necessity.

Summary

There appear to be good reasons for dividing the migratory patterns of marine species into several categories. This, however, presents major problems of definition which, unless resolved, are likely to plague future arrangements for fisheries. The definitions need to be made as precise as possible. It should be one of the tasks of the UN Conference to attempt to improve definitions if, as appears desirable, the concept of separate treatment is accepted. But in addition, it may be necessary to adopt a species-by-species approach to the problem of definition, explicitly placing each presently utilized species into one or another of the four categories. This would be particularly desirable for the categories of anadromous/catadromous species and highly migratory oceanic species. For this task, it may be desirable to establish an international body. It would seem that the task of preparing such lists would be too difficult and time-consuming to be undertaken at the UN Conference.

This paper deals primarily with the last of the four categories—that of the coastal species. As stated above, there is another Study Paper in this series dealing specifically with tuna and with the kinds of separate treatment that might be accorded to that member of the highly migratory oceanic category. The problems of the anadromous species relate, at the moment, primarily to salmon and are of interest to relatively few states. Sedentary species can generally be considered as a subclass of coastal species, usually subject to the unilateral control of a single coastal state. For these reasons, major emphasis is given in this paper to the category of coastal species and, within that class, to the situations where stocks are found within the waters of two or more neighboring states and within and outside coastal state jurisdictions.

MAXIMUM SUSTAINABLE YIELDS FROM WILD STOCKS

There are a few situations in which cultivation practices may improve yields of organisms, but these situations are restricted to a very small number of high unit-value species that meet special conditions. The vast majority of marine species that are of value to man are not economically susceptible to conservation practices. The yields from these resources are thus limited by natural conditions over which man has little or no control. For any particular stock of fish, there is a maximum annual yield that can be sustained over time. Employment of fishing effort beyond that point will generally lead to lower sustainable yields in the future, either because the reproductive capacity of the stock has been significantly diminished or because the tonnage of fish removed is greater than the tonnage that can be added by natural growth. When this occurs, the stock is said to be depleted.

The nature of depletion can be shown in the illustrative yield curve in Figure 1, relating annual sustainable yields to the amounts of fishing effort employed. Fishing effort is here measured by the number of standardized vessels times the number of days each vessel spends in fishing for the stock.

The curve TR, TY shows that, at low levels of effort, annual sustainable yields are also low. The yields are greater at greater amounts of effort, but at a declining rate, and there is a maximum yield that

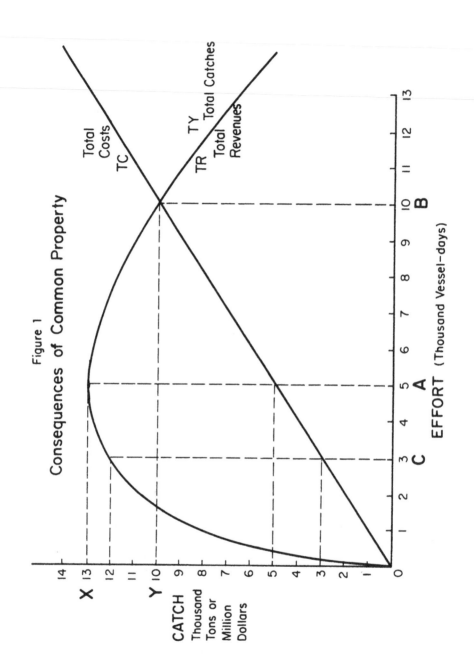

Figure 1

Consequences of Common Property

EFFORT (Thousand Vessel-days)

the stock is capable of producing on a sustained basis. In the illustration, the maximum sustainable yield is shown at point X, where 13,000 tons of fish are caught and 5,000 vessel-days are employed.

If greater amounts of effort are used, beyond the point of maximum sustainable yield, the annual catches from the stock are actually lower than they could be, and depletion has taken effect. For example, if 10,000 vessel-days are employed (twice the amount at the point of maximum sustainable yield), the annual catch is only 10,000 tons of fish (as against the 13,000 that could be produced).

It should be noted that there is a wide variety of yield curves, varying in the rate with which they rise and fall and in the sharpness of the peaks. For some species of fish, high levels of catch may be maintained over a wide range of amounts of effort, while for others, the yield may peak sharply and fall rapidly. Depletion may be more significant in the latter case, but economic waste (as noted below) may be great in both cases.

The number of depleted fish stocks is large and growing rapidly. A recent report by the Food and Agriculture Organization of the United Nations stated that, at the time of the first UN Law of the Sea Conference in 1958,

> ... the heavily exploited stocks were limited to some half dozen stocks in the North Atlantic and North Pacific, mostly of the large, valuable and long-lived species (halibut, salmon, plaice or haddock), and the blue whales in the Antarctic . . . [Now] the number of heavily exploited stocks has greatly increased, and includes stocks in all parts of the world. Fortunately only a few stocks have been so depleted that the present yield is only a small proportion of the possible yield . . . . Many more stocks have been sufficiently depleted for the effort expended in harvesting them to be much in excess of what is necessary, often with some decline in the total yield. In others, the effort is now increasing, or is likely to increase in the near future.[3]

In the absence of appropriate controls, depletion is inevitable as long as the demand for the stock continues to increase. It is very important to keep this point in mind and to avoid being misled by the apparent abundance of unutilized or underutilized stocks, or by opportunities emerging from changes in fisheries jurisdiction. There are a number of situations where coastal states can increase their catches by significant amounts (even though in aggregate, these may not add up to much relative to total world catch). Such situations

---

[3]FAO UN *Conservation Problems with Special Reference to New Technology,* Fisheries Circular No. 139 (Rome: FAO, 1972), p. 6.

11

occur where new markets can be developed, where new techniques for fishing or new fishing effort can be employed, or where foreign fishing effort has been displaced by coastal state extension of jurisdiction.

In these situations, an increase in coastal state fishing effort may be quite desirable. But, if the increase in effort is not properly managed, it will eventually and inevitably lead to depletion of the stocks and to economic waste. And once this has occurred, the establishment of appropriate controls becomes extremely difficult. Indeed, it is generally true that the greater the amounts of biological and economic waste, the more difficult it becomes to impose satisfactory controls. Thus, it is of critical importance to initiate controls before depletion takes place and to avoid excessive encouragement of fishing effort. It is also important, as noted below, to adopt those kinds of control that will be the most effective in social and economic terms.

INTERRELATIONSHIPS IN FISHERIES

Another characteristic of fisheries that creates special difficulties for management lies in the frequently complex interrelationships among different kinds of fishing effort. Three might be mentioned: (a) the intermixture of species; (b) predator-prey and competitive relationships; and (c) conflicts between different kinds of fishing gear.

The first of these—the intermixture of species—occurs on many grounds throughout the world and is often of such a nature that the fishing gear cannot select out the desired species. For example, on Georges Bank in the Northwest Atlantic, some fishermen fish only for cod, but their trawls bring up large incidental catches of haddock and other groundfish. It has been estimated, in this particular situation, that the incidental catch of haddock is greater than the maximum catch that should be taken if the haddock stock is to be rehabilitated.[4] While this may be an extreme case, the problem of achieving optimum yields of all stocks taken in the same nets is a common one throughout the world.

Ecological interrelationships between species are also important and present difficulties for management. In several situations, closely related species compete for the same biological niche—they occupy the same waters and feed on the same materials. In these cases, the fishing of one species may lead to its replacement by the other. An extreme example of this can be found in the Great Lakes of North America, where several different kinds of fish have become succes-

[4]"Memorandum by the U.S. Commissioners on the Regulation of Fishing Effort, Advance Copy," 14 November 1972, p. 4.

12

sively dominant due to high rates of fishing and to changes in natural mortality. Another example can be found in the shift from sardines to anchovies off the coast of California. In addition to competitive interrelationships, difficulties sometimes occur for species involved in predator-prey relationships. If the population of (and the yield from) a predator is significantly reduced, it may permit higher than normal yields from the prey. A significant reduction in the population of the prey species, however, may mean lower populations and yields from the predator, or perhaps a shift to another prey species.

Such ecological interrelationships are very complex and not well understood. At present, it is not known how important they are in terms of man's interests, but as more species of fish are brought into utilization the chances of interferences between ecologically related stocks become greater.

Conflicts between different kinds of fishing gear, however, are already known to be of considerable importance in a number of areas. These conflicts occur when sweeping gear, such as trawls, are used in the same area as stationary gear, such as lobster or crab pots or longlines.

In each of these kinds of interrelationships, there are problems in determining the appropriate levels of catch for the species. The yields from each of the interrelated species cannot be maximized simultaneously because the maximum yield of one will mean less than the maximum yield of the other through either over- or underutilization. This provides one example of the difficulty of using the conventional goal of maximum sustainable yield as a basis for making decisions.

Where these interrelationships exist, a choice has to be made on the best combination of yields from the stocks. It would be of little value to man to base the choice on the combination that produces the greatest tonnage of fish because this might mean a larger production of the low-priced species and smaller production of the high-priced species. For example, in a predator-prey relationship, the maximum tonnage might be obtained by a severe reduction in the population of the predator species and the achievement of a high yield from the prey species. But if the prey species is used for fishmeal and the predator is a high-quality product consumed in a luxury market, the greatest economic return is likely to occur under the opposite conditions.

Economic values are clearly preferable to physical yields as a basis for making choices in such situations. Unfortunately, however, the application of economic values is often very difficult and sometimes in conflict with other benefits sought from fisheries. The states that share the related species may place quite different values on them. For example, in the Northwest Atlantic the fishermen from Spain place a much higher value on cod than on haddock, while the U.S.

fishermen prefer haddock over cod. In the absence of free trade of the commodities, the maximization of net economic revenue depends upon whose system of values is being used. But even with free trade, economic maximization is difficult to adopt because it may mean a significant change in patterns of catch and distribution of wealth.

Such conflicts cannot be easily resolved. But, as in other situations, resolution can be facilitated by better definition of the benefits that are sought, better understanding of the effects on these benefits of alternative solutions, and improved means for achieving acceptable trades in benefits. These elements, referring to distribution of wealth, are discussed more fully below.

ECONOMIC CONSEQUENCES OF COMMON PROPERTY

In addition to the biological characteristics that make fishery problems different and difficult, there is an important legal characteristic that should be mentioned. This is the traditional and widespread treatment of fishery resources as common property resources. Under this treatment, there is free and open access to the resources and no means for preventing simultaneous use by two or more users.

In national waters, the condition of common property may or may not exist depending upon the way in which the state chooses to manage its fisheries. In the waters of the Soviet Union and Japan the condition of free and open access has been removed and the use of each fishery is subject to the control of a single economic unit. In the waters of the North American and West European states, however, such controls do not generally exist. Although there are exceptions, these states have tended to maintain free and open access and to let as many economic units fish as desire to do so.

In international waters, the condition is universal, and derives from the principle of the freedom of the seas, which guarantees every state free and equal opportunity of access to all fishery resources. There is no legal authority by which a state or group of states can exclude other states from using a resource on the high seas (although there are instances where *de facto* exclusion has occurred).

The condition of free and open access is accompanied by severely detrimental economic consequences wherever it occurs. This can best be illustrated by referring to Figure 1, in which it was shown that the annual sustainable yield increases with increasing amounts of effort up to a maximum point and then declines because of depletion of the stock. This yield curve (TR,TY) can also refer to total economic revenues from the stock by applying a price to the annual sustainable catches. It is assumed in the illustration that the price for a ton of

14

fish is $1,000, so that if 10,000 tons are caught the total revenue from the fishery is $10 million.

It is also assumed that each day of fishing by each vessel costs $1,000. This includes not only the operating costs but also the normal returns expected for capital and labor (whether under a market or centrally planned economy). As illustrated by line TC (total cost curve), where 10,000 vessel-days are spent (say 50 vessels operating for 200 days of a season), the total costs of the fishery are $10 million.[5]

In a common property fishery, where there are no controls on the number of vessels, the fishery will operate at the point where the total costs of the fishermen are equal to the total revenues received. In the hypothetical illustration, this occurs at point B, where there are 10,000 vessel-days used during a season and the catch is 10,000 tons. Total costs and total revenues are both equal to $10 million at this point.

At any amount of effort less than 10,000 days (e.g., where there are fewer fishermen), the total costs of the fishermen are less than the total revenues received, and there is an extra profit available to the fishermen. For example, at point A, there are only 5,000 vessel-days of effort, but the catch is higher and amounts to 13,000 tons—so that the total costs are only $5 million while total revenues are $13 million. At this point (which is also the point of maximum sustainable yield) there is an extra profit of $8 million.

Where such a profit exists (as in the development of a new fishery) it obviously makes fishing a very attractive proposition. And since there are no controls over entry into the fishery, the extra profit will attract more and more fishermen. Thus, more and more fishermen will enter, even though they add to the total costs of the whole fishery and simultaneously reduce the levels of catch and the total revenues. In the usual case, fishermen will enter until all extra profit disappears and total costs become equal to the total revenues (at point B in Figure 1).

As noted above, the figure has been drawn to show that the waste occurs in a physical sense as well as in an economic sense; that is, it shows that the fishery is depleted, producing only 10,000 tons of fish instead of the 13,000 tons that is the maximum yield that can be sustained over time. While this is not an unusual situation in the real world, it should be pointed out that the physical consequences

[5]The assumption that both price and unit costs are constant serves to simplify the illustration. In fact, prices may vary with regard to amount of production thrown on the market and size of fish, and costs are likely to vary with regard to amount and kind of effort. Allowing for the variations, however, will not change the results illustrated.

may be quite different. The total cost curve could be drawn to intersect the total yield and revenue curve at the point of maximum sustainable yield, or below that where the fishery is often (and incorrectly) said to be underutilized. However, even in such cases, economic waste will still be present because of the condition of free and open access.

In the illustration, the hypothetical fishery tends to operate at point B. The economic efficiency would be greatly improved by the use of less fishing effort. If effort could be controlled at point A, the greatest catch would occur and there would be an extra profit of $8 million that could go either to the fishermen, to society, or both.

However, it should be noted that an even larger extra profit could be produced at an even lower amount of effort. At point C the total costs are only $3 million while total revenues are $12 million, and there is an extra profit or economic rent of $9 million. It is at this point that most natural resources industries would tend to operate, either under free market or centrally planned economies. Since most industries, unlike fisheries, can control the entry of capital and labor, they do so at the point of greatest net return. It would be economically wasteful to move from point C to point A, since the additional $2 million of capital and labor would only produce an additional $1 million of revenues. In most circumstances, the extra capital and labor could be employed more beneficially in other forms of activity.

Thus, in purely economic terms (and assuming constant prices), the most efficient operation of the fishery would occur where the maximum net economic revenue is produced rather than at the point of maximum sustainable yield. However, it should be noted that the point of maximum sustainable yield is so thoroughly ingrained in fishery management practices that it may be extremely difficult to adopt a lower yield as a goal even though it would produce greater economic returns. But even if this is the case, efficiency can still be greatly increased by controlling the amount of fishing effort.

EXAMPLES OF ECONOMIC WASTE

While the above discussion is theoretical and somewhat oversimplified, the fact of economic waste has been amply demonstrated in a number of studies of particular fisheries. In the Pacific salmon fishery of the United States and Canada, it has been estimated that the same annual catch (and total revenue) could be taken with about $50 million less capital and labor than are currently employed each year.[6] And

[6]James Crutchfield and Giulio Pontecorvo, *The Pacific Salmon Fisheries: A Study of Irrational Conservation,* (Baltimore: Johns Hopkins Press for Resources for the Future, Inc., 1969), p. 174.

16

this estimate assumes that the present highly inefficient and wasteful conservation measures continue to be enforced. Peru, although catching the largest quantity of fish of any country in the world, is also suffering economic waste because of overcapitalization. It has been estimated that the excess capacity in fishing vessels and in processing plants is so great that savings of $50 million a year could be achieved by reductions in effort.[7] In 1965, the cod fishery of the North Atlantic was so overutilized that the same amount of catch could have been taken with 10 to 20 percent less fishing effort than was employed. Such a reduction would have produced savings estimated at $50–100 million per year.[8] Since there has been a considerable increase in effort since 1965 (with no increase in total catch), the savings would be far greater today. In the area of Georges Bank in the Northwest Atlantic (Subareas 5 and 6 of the ICNAF region), total fishing effort in 1971 was estimated to be 31 percent more than that required to achieve the maximum sustainable yield from the groundfish stocks being used.[9] In the first seven months of 1972, total effort was 35 percent greater than during the same period in 1971.

There are no estimates of the total world losses due to the application of wasteful amounts of effort. They are, however, likely to be extremely large and amount to $1 billion or more per year. This does not mean that such an amount could be readily extracted for the benefit of society. And it says nothing about the costs and difficulties of adopting management techniques that would permit such extraction of rents, nor the difficulties of determining who should divide the rents.

Many of the developments taking place in fisheries tend to aggravate the wastes rather than alleviate them. For example, the growth in demand and the development of certain technological innovations tend to make waste more, rather than less, severe. This is also true of most traditional conservation regulations.

For the world as a whole, the demand for fisheries is continuing to increase, due to increasing numbers of people and to higher per capita incomes. But since demand is increasing more rapidly than the supply of fish, the prices are also increasing, producing higher revenues to the fishermen, and, thereby (temporary) extra profits. But

[7]L. K. Boerema and J. A. Gulland, "Stock Assessment of the Peruvian Anchovy and Management of the Fishery," paper FI: FMD/73/R-6, February 1973, presented at the FAO Technical Conference on Fishery Management and Development (Vancouver, B.C.), p. 9.

[8]"Report of the Working Group on Joint Biological and Economic Assessment of Conservation Actions," International Commission for the Northwest Atlantic Fisheries, Committee Document 67/19, Annual Meeting, June 1967, p. 4.

[9]"Memorandum by the U.S. Commissioners," supra note 4, p. 12.

since access is free and open, the extra profit will attract more fishermen and lead to higher total costs to match the higher total revenues. Thus, higher prices for limited supplies attract more fishermen and aggravate economic waste.

Similar consequences will result from certain technological innovations. If these innovations reduce the costs of catching an already fully utilized stock or increase the catches per vessel-day, an extra profit will appear. Depending upon the nature of the innovation, this could eventually lead to greater or lesser amounts of effort.[10] But in either case, the extra profit will disappear, total costs and revenues will become equal, and stock will tend to become further depleted. It should be noted that other technological innovations might not have such a deleterious effect, since they might make available new stocks that cannot currently be fished or improve selectivity of gear and, thereby, reduce incidental catches.

In the absence of appropriate controls, economic waste is just as inevitable as biological waste. States contemplating expansion of fisheries should be thoroughly aware that contributions to the growth of their economies may be severely curtailed by failure to adopt economic controls. Where there are opportunities for expansion of catch (because of new markets, new sources of supply, new fishing techniques or gear), there is a natural temptation to encourage investment in capital and labor. Such investments may be quite profitable —during their initial stages. But profitability, while desirable for the economy, also carries with it the seeds of its own destruction. It will attract more and more investment, putting greater pressure on limited supplies. Returns to capital and labor will decrease and the extra profits or economic rents that could have been used for other development activities will be dissipated. Thus, contributions to the economy by

[10]If the innovation reduces the cost of a day's fishing (by reductions in amount of labor required or use of cheaper fuels, for example), the initial result would mean that the same number of fishermen are catching the same amount of fish, but at a lower cost. More fishermen will enter, because of the extra profit available, and add to the total costs of the fishery. If the stock is already depleted, depletion will become more severe.

If the innovation leads to higher catches per vessel-day (by the use of better nets, for example), the initial result will be higher total catches and revenues for the same number of fishermen. This will also attract more fishermen, but the catches will not be sustainable and could have a severely depleting effect. As the effect takes place, the lower total catches and revenues will force some of the fishermen out of the fishery and could lead to lower total employment in the long run.

18

the fish catching industry will become negligible unless satisfactory controls are established.[11]

The only kinds of controls that can prevent such economic waste are those that remove the condition of free and open access and that limit the amount of fishing effort. This, however, is extremely difficult to do in many situations, because it means a limit to the opportunities for employment. If alternative employment opportunities are not fully available, a state may prefer to maintain the conditions of free and open access even though returns to labor are low and economic waste is great. But before making such a decision, a state should be aware not only of the income that it will be sacrificing but also of the costs it is likely to bear. Management of free and open access fisheries tends to be extremely difficult and generally leads to severe inhibitions on technological innovations. The net result is likely to mean that the additional opportunities for employment are achieved at great cost and through a considerable loss of potential benefits.

The difficulties of removing the condition of free and open access are by no means small where stocks are fully enclosed within the waters of a single coastal state. But where stocks swim in the waters of two or more states, the difficulties are greatly compounded. No individual state will voluntarily and unilaterally reduce or limit its fishing effort on a shared stock, because any gap it leaves will be taken up by others. Unified control is essential but the adoption of controls that permit optimization of economic yields are made difficult because they directly involve the distribution of fisheries wealth; that is, the condition of free and open access cannot be removed without making explicit decisions about allocation of, access to, or revenues from the resource. Such decisions are inevitable in the long run, no matter what values are sought from fisheries, but their inevitability does not reduce their difficulty.

CONFUSION BETWEEN PRODUCTION AND DISTRIBUTION

One of the reasons for the difficulty in making decisions lies in the confusion between means for the production of fisheries wealth and means for its distribution. This confusion is an important characteristic of marine fisheries and the consequence of the principle of the freedom of the seas. That principle guarantees that all states shall

---

[11]An excellent discussion of several aspects of fisheries and economic development can be found in Anthony D. Scott, "Fisheries Development and National Economic Development," *Proceedings of the Gulf and Caribbean Fisheries Institute* (18th Annual Session, November 1965).

have equal access to fisheries beyond the limits of national jurisdiction. But it does nothing more than that. Thus, only those states that exercise their right of access gain in the distribution of the wealth of the fisheries. Those states that do not exercise their rights on the high seas receive nothing.

This condition does not generally exist for publicly owned natural resources that fall within national jurisdictions. Conventionally, the state extracts some degree of wealth in some form from the users of public resources and the wealth is then shared by society as a whole. While there may be many differences (and imperfections) in the ways in which this is accomplished, there exists, at least, the principle by which those that do not actually have access to the resource gain benefits from the resource users. For example, in the case of oil resources on the U.S. continental shelf, society receives considerable benefits from the lease fees and royalties paid by the exploiters. And for fisheries within national jurisdictions, the distinction is made when the coastal state extracts wealth by charging fees to the foreign fishermen that use the resource.

But on the high seas, there are only two instances where a distinction has been made between the right of access to the resource and a right to share in its wealth. One of these is the arrangement for the fur seals of the North Pacific, which provides that the seals shall be harvested only by those states on whose islands the seals breed. This allows for the most efficient harvesting and avoids the wasteful practices of seeking out the seals when they are dispersed on the high seas. Thus, the maximum net values can be produced. The states that formerly harvested seals have agreed to abstain from doing so and, in return, they receive a share of the skins.

A less direct instance of the distinction can be found in the agreements for the Antarctic whales. In these agreements, several states acquired national quotas permitting them to harvest certain numbers of whales. The quotas were transferable and those states that sold them gave up their right of access and received, in return, a share of the wealth.

But these instances are unique, and the prevailing general principle is that the high seas wealth in fisheries is only shared by those who continue to exercise their right to fish. Two detrimental consequences emerge from the maintenance of this principle. One is the confusion between objectives of management and distribution and the other is the constraint that the principle imposes on the range of alternative solutions.

The objective of management is to produce the maximum net benefits from the resource, irrespective of who acquires the benefits or how they are distributed. The objective of distribution is to produce a sharing of the wealth in a manner that will be acceptable to those states that have an interest in the resource. If the means for the production and acquisition of wealth are inextricably related, as they are now, states will attempt to acquire wealth by the use of management techniques or arguments. In some cases, it is stated that a country that is best situated to manage a resource should have exclusive control over access and, therefore, exclusive rights to the wealth. In other cases, management measures may be proposed that discriminate against some of the present users or that impede access to new users. The failure to make distinctions between production and distribution adds to the difficulties of negotiation. It also serves to reduce the range of choice among alternative solutions because it means that states will be unwilling to accept arrangements that reduce their access to the resource. For example, in many situations the maximum net benefits might be produced by permitting a single state to have sole access to the resource. But this solution is unacceptable as long as the other states that give up the right of access get nothing in return. If, however, a distinction is made between production and distribution, then it becomes possible to adopt a wider range of solutions.

SUMMARY

The fundamental characteristics of fisheries, described above, distinguish the industry from most other forms of enterprise and activity. They also indicate some of the complexities of management and the challenges to be found in the search for, and adoption of, new regimes.

Some of the problems derive from the physical nature of the resource. The fugacity of the fish and their disrespect for man-made boundaries means that unified control is only possible through agreement among the relevant states. The inability to cultivate most marine organisms means that supply is limited by natural conditions and cannot be increased beyond a certain point no matter how much capital and labor are invested.

Other problems derive from the legal heritage of the principle of the freedom of the seas; i.e., the condition of common property under which there is no control over access and no satisfactory rights of property. The future has no value, for anything left in the sea for tomorrow will be taken by others today. Under these conditions, both biological and economic waste are inevitable unless appropriate controls are adopted. The adoption of appropriate controls, however, is frustrated both by the natural conditions of the resource and by the traditional confusion between the objectives of management and those of distribution.

II Alternatives for Management

MANAGEMENT GOALS

STATES, in seeking better arrangements for shared fisheries, face a wide assortment of alternative techniques for management. These alternatives include techniques that may be related primarily to the resource itself, or to the kind or amount of fishing effort. They may range from a minimum of controls (with access to the resources both free and open) to almost absolute control over both the quantity and kind of use. With such a variety of alternatives, the choices may be very difficult. This task, however, can be facilitated by evaluating the different alternatives in terms of the various goals to be sought from the management of marine fisheries.

The many and varied goals sought from the use of fisheries differ widely in their relative importance to different states; in the degree of their validity in terms of man's social and economic interests; and in the degree to which they may be mutually complementary or conflicting. Not all of the many goals can be discussed in this paper. Instead, discussion is restricted to three major goals that are considered to be of particular importance for joint decisions by states and for their implications for different management techniques. The first of these—the optimization of biological yields—relates primarily to the resource itself. The second—the optimization of economic and social benefits—relates to the users of the resource. And the third—the reduction of management costs—relates to the organization governing the resource and resource users. All three of these goals must be considered in conjunction when evaluating the different techniques for management.

It should be recognized that the choice of arrangements may be strongly influenced by other goals than the three discussed, as well as by interests totally unrelated to fisheries. Nationalism, fear of foreigners, desires to maintain the freedom of the seas as a means for ensuring mobility of naval vessels, and a host of other nonfishery-related goals will obviously affect a state's views toward alternative fishery arrangements. However, it would be both impossible and inappropriate for this paper to deal with such objectives. All that can be

22

said in this regard is that states should carefully examine the sacrifices in fishery benefits that might be incurred in reaching these other goals.

Optimum Biological Yields

One of the goals for fisheries management is that of optimizing the sustainable yield from the resource. It is often stated in international conventions and agreements that the primary objective is that of maximizing (or optimizing) sustainable yields from the stocks. The maximum sustainable yield is achieved when the annual catch is at the highest level that can be sustained over time. In order to achieve this goal, it is generally necessary to reduce current levels of catch so that the stock can be rehabilitated. After rehabilitation, it becomes necessary to control the catch at the level appropriate to the factors of natural mortality and recruitment. Optimum sustainable yield generally means the same thing, except that this goal permits the achievement of something less than the maximum where the use of the stock is interrelated with the use of other stocks in such a way that yields cannot be maximized simultaneously.

In large part, the importance of such resource-related goals is due to the long history of their use in international agreements and conventions. But the goals of maximum or optimum sustainable yields are, by themselves, of little relevance to the interests of man. They refer essentially to physical quantities and are generally adopted without regard to the associated costs and benefits. The questionable validity of such goals can be illustrated by applying them to the use of terrestrial resources. States do not accept as valid the goal of maximizing the sustainable yield from an acre of ground. If they were to do so, they might be producing several times as many bushels of wheat, rice, or corn per acre as are now being produced. But this could only be done by incurring costs that are much greater than the revenues achieved, or by diverting scarce capital or labor away from other more profitable or productive activities. Similarly, it makes little sense for states to maximize sustainable yields from fisheries without regard to the costs and revenues associated with the production.

In spite of this, a resource-related goal does have some value and should be included in the consideration of alternative regulatory techniques. Management techniques that optimize biological yields may, for example, be better than no management at all. And in some cases, acceptable economic and social goals may be so immediate in nature or limited in scope that the imposition of a resource-related goal may be desirable for the long run.

Although the goal of optimizing biological yields cannot be defined with precision, it can be characterized by discussing various situations in which it might be employed. For example, the utilization of a resource to the point of extinction would be clearly damaging to the interests of the world community as a whole. It may, therefore, be desirable for the delegates at the UN Conference on the Law of the Sea to adopt a general principle prohibiting any state or group of states from using a resource to the point of extinction. This might be further elaborated by providing for the imposition of international controls in a situation where a state is unwilling or unable to prevent extinction of a stock that falls within its zone of fishery jurisdiction.

It is less easy to define the goal of optimizing biological yields in terms of depletion than it is in terms of extinction. Depletion is a clear manifestation of waste. But the waste is of biological materials and not necessarily of economic or social values. In some situations, countries may be impatient, and set higher values on today's output than on yields they can expect in the future. For example, a rapid exploitation and depletion of a stock may provide them with capital that they can invest in other, more productive enterprises. Or they may anticipate a decline in the relative price of the marine product. For these countries, sustainable yields would have little appeal.[1] If this attitude applies to a stock that falls entirely within the jurisdiction of a single coastal state, and if the stock is not threatened with extinction, then questions can be raised as to whether a universal principle against depletion is desirable or will be acceptable at the UN Conference. Those in favor of such a principle may feel that the oceans' resources, particularly those within extended zones of jurisdiction, are part of the common heritage of mankind. They may feel that no individual state has a right to "misuse" those resources because of its short-term or narrow interests, or that no state has a right to diminish the world's food supply even temporarily. Those opposed to a general principle prohibiting depletion may point out that such principles do not apply to a state's forest, agricultural, or other renewable natural resources found on land and should not, therefore, apply to its fisheries. They might point out that a right to deplete a resource that lies wholly within a state's borders is a matter for that state's decision and not an appropriate subject for international controls.

Another situation in which the goal of optimum biological yield might or might not be invoked occurs where a stock is shared by two or more states whose economic and social values differ considerably. The conflicts between their values may be so great that the states find it difficult, or perhaps impossible, to agree among themselves

---

[1]Christy & Scott, *The Common Wealth in Ocean Fisheries* (Baltimore: Johns Hopkins Press for Resources for the Future, Inc., 1965), p. 218.

24

on appropriate management measures. This may lead to a severe depletion of the stock and a loss of benefits to all parties. In this situation, it might be proposed that the states must accept the imposition by an outside agency of controls designed to maintain or enhance biological yields. The agency might continue to impose such controls until the states settle their differences either on their own or through compulsory arbitration. This principle would be designed to prevent mutually destructive patterns of behavior and to provide states with at least a minimum of benefits until they are able to adopt better arrangements. It is questionable whether this could be adopted as a universal principle at the UN Conference. It might be more appropriate for adoption in specific regional arrangements.

The goal of optimizing biological yields might also be of value in situations where stocks are so closely interrelated that the use of one has a significant effect on the use of another. This kind of situation presents special difficulties for management that cannot be readily resolved where the economic and social interests in the related stocks are different. If it is likely that agreement cannot be reached, it may be desirable to limit the catch levels of the stocks so that eventual rehabilitation can be permitted without too much difficulty.

The adoption of the goal of optimizing biological yields would not be necessary if states had uniform economic and social values and uniform views on future returns. But since uniformity does not exist, it may become necessary to impose limits on the levels of catch, and to insist that yields not be diminished beyond a certain point. Aside from general agreement that levels of catch should not be so great that they lead to the extinction of a species, it is impossible to determine the amount of biological yield that would be the optimum. Where economic and social values are not introduced, the choice of levels of catch is an arbitrary one. But even so, it may sometimes be preferable to adopt an arbitrary limit than none at all.

Optimum Economic and Social Benefits

It would appear obviously desirable for states to adopt the goal of optimizing economic and social yields from fisheries. But for various reasons, this goal has never been adopted in international agreements or conventions, and has received little, if any, attention. One of the reasons for this is that the enhancement of economic yields from over-utilized fisheries cannot be achieved unless the states deal directly with the controversial problems of wealth distribution; that is, net economic returns can only be achieved by controls on the amount of fishing effort, and this can only be done by allocating the resource, access to the resource, or resource benefits. The unwillingness to face up to the problems of distribution has impeded the adoption of economically desirable goals in all but a very few special cases. However, this situation has now changed, since it has become quite clear that wealth distribution is a fundamental problem of fisheries that can no longer be avoided.

A more persistent and difficult problem is that definitions of economic and social values may differ quite widely and that different values may be in conflict. One particular conflict creates special difficulties in the search for improved arrangements: the conflict between the goal of increasing net economic returns from a fishery and that of increasing opportunities for fisheries employment. As discussed in Chapter I, an increase in net economic revenues from a fully utilized fishery generally necessitates a reduction in numbers of fishermen. Thus, these goals cannot be maximized simultaneously, and the achievement of one requires some sacrifice of the other.

In this paper, the economic aspects receive relatively more attention and value than the employment aspects. There are several reasons for this emphasis. One is simply the fact that employment characteristics of fisheries are generally well understood but, in many countries, economic characteristics have frequently been neglected. For example, even in such a developed market-oriented country as the United States, fishery administrators have often ignored the economic consequences of the management practices they have adopted.

A second reason for emphasizing economic goals is that they are believed to be of more value to a nation than the opportunities for employment that are provided by fisheries. In many fully utilized fisheries, the average income levels tend to be quite low—frequently lower than those found in similar natural resource industries. This is due not only to the condition of free and open access that leads to the dissipation of economic rents in the industry, but also to the fact that it is generally easier to enter a fishery than it is to leave it. Many fisheries are subject to natural fluctuations in yields, providing high levels of catch in certain years and low levels in others. Vessels attracted into the fishery during the good years tend to remain during the bad years because there are few, if any, other opportunities for their use. Because of the fluctuations, labor and capital may persist in a fishery even during the years when the returns are unsatisfactory.

The social gains from fisheries employment may, therefore, be quite small where the condition of free and open access is maintained. The general interests of the country might well be better served by adopting controls that prevent superfluous capital and labor from entering fisheries that are already fully utilized. Depending upon the nature of the controls, this could lead to higher returns to capital and labor or to greater economic revenues to society (or to a combination of both). Such controls may be politically difficult to impose, but once established they are likely to produce greater benefits to a state than the maintenance of low-income labor in a depressed industry.

An additional reason for giving preference to economic values lies in the fact that the controls tend to be less restrictive and growth-inhibiting than those used where the objective is to increase employment opportunities. This is discussed more fully later, but it can be pointed out here that, if free and open access is maintained, the prevention of depletion tends to require prohibitions against technological innovations and increasingly inflexible restrictions on gear and vessels. For example, in the Maryland waters of the Chesapeake Bay in the United States, employment in oystering is maintained by the requirement that oyster dredging must be done by sailboats.

Finally, the problems of management and distribution are easier to resolve where states choose economic objectives rather than employment benefits. As noted previously, it is easier to distribute fisheries wealth in the form of economic revenues than in the form of access to the resource. And where this is acceptable, there are also fewer constraints on the choice of management techniques.

These various reasons indicate the author's preference for the goal of optimizing economic yields rather than employment opportunities. However, the choice is clearly a matter for individual states and for states acting together in the joint management of a common resource. Since there are likely to be differences in the ways in which states view the different benefits, no precise definition can be given to the goal of optimizing economic and social yields. The most that can be done in this paper is to illustrate how the different benefits will be affected by the alternative management techniques.

Reduction of the Costs of Management

In most natural resource industries, management decisions fall naturally into the hands of the entrepreneur or agency that has exclusive rights to the resource. But in shared fisheries, no individual user is capable of managing the resource by himself. Instead, a separate agency must be established and must be assigned the powers necessary to make the management decisions. This separation of the objectives and functions of the managers from those of the users places a unique and heavy burden on management. In addition, this separation means that the costs of management are not generally borne by the users. They are thus treated as external to the costs and revenues of the industry, and are not included in the calculations of profitability. For example, when it is stated that the same total catch and total revenues could be taken from North Atlantic cod with $50–100 million less capital and labor than is employed per year, this overestimates the potential returns because it ignores the costs of research, administration, negotiation, and enforcement.

Thus, a critical, but generally neglected, goal for fisheries is that of reducing the costs of management. The costs can vary considerably depending upon the management techniques and regulatory measures that are chosen. For example, if a total limit on annual catch is adopted as the technique of management, it requires sufficient knowledge to determine that limit. But if the fishery is regulated by controls on fishing effort, it is necessary to acquire information on the fishing power of the different kinds of vessels and gear, in addition to information on the yield from the stock.

With regard to enforcement, there are several management techniques that can only be enforced through inspection of the gear that is used or the catch taken, requiring a large number of enforcement vessels. Other management techniques might require little more than a count of the number of vessels, which could be achieved at relatively low cost through air surveillance.

Another example can be found in the different techniques for extracting revenues from foreign fishermen. One technique is to impose a tax or royalty on the amount of catch, requiring satisfactory knowledge about the amount of catch taken by each vessel during each trip. It would be less costly to extract revenues by the imposition of license fees for each vessel. Even less expensive would be a technique that leases rights to all fishermen of a foreign state. In choosing among these different alternatives, the coastal state should balance the costs of collecting revenues against the effectiveness of the techniques.

In addition to research and enforcement, the management agency should also consider the costs of negotiation. These tend to increase with the complexity and inflexibility of the regulatory measures. The use of national quotas, for example, requires annual determination of the appropriate total limit on the catch of each stock as well as on the distribution of shares among the member states. Controls on fishing effort through direct limits on the number and kind of vessels are even more difficult. It requires negotiations to determine the meaning of fishing effort and differences in effectiveness of different size vessels and different equipment.

In the past, the choice of management techniques has sometimes been made with little or no consideration of the costs of research, administration, enforcement, and negotiation. This may have had a detrimental effect on management because of inability or unwillingness to follow through with sufficient funds or expertise. It might have been preferable to choose a more modest technique that could have been managed at less cost. In short, decisions on alternatives should not be based solely on their ability to reach the goals of optimum biological yields or optimum economic and social benefits, but also on the costs that will be incurred in management.

ALTERNATIVE REGULATORY TECHNIQUES

The functions of management include those of research, regulation, and enforcement. Of these, the problems of regulation are the most important, both because of their effect on research and enforcement efforts and because of their direct relationship to the goals of management. In addition, the problems of regulation are the most difficult to comprehend. There is a wide variety of techniques that can be used to regulate fisheries, whereas the techniques for research and enforcement are fairly simple and clear-cut. The problems of regulation, therefore, receive primary consideration in this chapter, while the problems of research and enforcement are discussed more fully in the consideration of alternative fishery institutions.

The many different techniques for regulating fisheries can be loosely divided into three categories: regulations relating directly to the resource; those governing the *kind* of fishing effort; and those controlling the *amount* of effort. Since certain techniques might be considered to fall within more than one category, the classification is not precise.

Resource-related regulations deal directly with the stock of fish, and include such techniques as closed seasons, closed areas, and limits on the total catch (including limits divided up into national quotas). Regulations governing the *kind* of fishing effort are generally known as gear restrictions. The techniques include controls over the size of mesh that can be used in nets, limits on the size of vessels, limits on the size and number of hooks, etc. Regulations on the *amount* of fishing effort have not generally been adopted in internationally shared fisheries, although there are a few minor exceptions. Controls on effort can be achieved directly, by limiting the number of vessels, or indirectly, by the imposition of license fees or taxes which operate as a disincentive for investment in fishing effort.

A significant difference between the various techniques lies in the degree to which they maintain the condition of free and open access to the stocks. With the exception of national quotas, the techniques in the first two classes mentioned above preserve this condition. However, for national quotas and for controls on the amount of fishing effort, the condition is removed—access is either limited or is made subject to some form of payment. Techniques that remove the condition of free and open access have a direct effect on the distribution of wealth, whereas those that preserve the condition do not. It is, therefore, far more difficult to adopt national quotas and controls on the amount of fishing effort than the other techniques mentioned above.

Resource-Related Regulations

The value of resource-related controls depends upon the specific purposes for which they are adopted. If the purpose is to deal with the problems of incidental catches and other kinds of interrelationships among species or among fishing gear, then closed seasons or closed areas may be quite desirable. For example, it may be desirable to prevent the use of sweeping gear in an area that has a particularly high value for the use of stationary gear. Or incidental catches may be more of a problem at certain times of the year than at others, in which case a closed season might be helpful.

However, if the purpose of the controls is to achieve or maintain a certain yield from a stock, the use of closed seasons or closed areas is generally undesirable in economic terms and tends to become ineffective in the long run. Closed seasons and closed areas can temporarily reduce levels of catch of an overexploited stock because the fishermen incur increased costs in having to go greater distances or concentrate their effort during periods when the stocks are dispersed. But if the demand for the product increases, as is likely, more fishermen will be willing to incur these costs, fishing effort will increase, and levels of catch will rise. If closed seasons and areas are the only means for controlling levels of catch, they will have to be continually enlarged as demand continues to increase. This is economically very damaging, since it raises the costs of fishing but not the quantity of fish that is harvested.

A limit on total catch is the most direct of all techniques for controlling yields from a stock. Under this technique, a total limit is determined prior to the opening of the season, and all fishing stops when this limit has been reached. But while it may be the most direct technique for achieving this physical goal, it has extremely damaging economic consequences.

Where a total quota is imposed, it creates a strong incentive to increase fishing effort. More, larger, and faster vessels will be built since each fisherman is in a race with all others to get as great a share of the total catch for himself before the limit is reached and the season closes. This increases not only the direct costs of the industry (with no increase in total revenues) but also the indirect costs. The excessive applications of effort mean that the limit will be reached sooner than before and that the season will shorten in length. In the case of the Pacific halibut fishery, the season dropped from nine months to four weeks in one regulatory area and to less than two months in the other area. In the case of the total quota for yellowfin tuna in the eastern tropical Pacific, the season has dropped to about three months from the usual nine months or more.

One consequence of this is that the catch reaches the market in a short period of time, placing strain on the processing, storage, and distribution facilities, frequently with detrimental effects on the quality of the product and the price to the fishermen. Another consequence is that the vessels must turn to other areas or stocks after the closing of the short season. This tends to place greater pressures on the other stocks. For example, after the season for yellowfin tuna closes in the eastern Pacific, many of the vessels move to the Atlantic and contribute to the excessive pressures on tuna in that ocean. The necessity for controls in the Atlantic leads, in turn, to further displacement of vessels and the eventual need for controls on a worldwide basis.

Thus, while a total quota provides the most direct means for preventing the depletion of a stock or for maintaining a desired yield, it also is one of the most damaging forms of control for man's economic interests. With regard to employment opportunities, the likely effect, as noted above, will be large numbers of fishermen working for only a short period of the year. If there are other fisheries available during the balance of the year, this result is not necessarily damaging. But if there are not, then it means that the superfluous fishermen have to turn to other activities or remain idle after the season closes.

The total quota system, however, can be significantly modifed by dividing the catch into national shares. This technique has been adopted in several situations—notably, for Antarctic whales, salmon, and crab fisheries between the Soviet Union and Japan, and, most recently, the groundfish fisheries of the Northwest Atlantic. This technique, which is essentially a distribution scheme, presumably permits the quota holders to decide for themselves how they wish to use their shares. Under ideal conditions, states would not feel compelled to invest large amounts of effort to capture the greatest share for themselves before the total limit is reached and the season closes. Each state, instead, could choose whether or not it wishes to use its quota to increase employment opportunities or to increase net economic revenues. This permits each state to operate according to its own goals and reduces the necessity for coming to agreement on common objectives.

Leaving aside the problems of distribution (discussed later), national quotas may greatly facilitate the resolution of conflicts over the use of shared stocks in many situations. However, the system will not necessarily prevent economic waste or enhance opportunities for employment. It may, for example, be less costly for a state to take its quota early in the season rather than late. This would occur where fishing effort disperses a stock or thins it, so that catches per haul of net decrease during the season. Under this condition, the incentive to invest large amounts of effort at the opening of the season would persist, and economic waste would continue. The fear of such results was, in large part, responsible for the recent U.S. proposal that the new national quota arrangement in the Northwest Atlantic be combined with limits on the amount of fishing effort. In spite of these difficulties, and assuming that the distribution problems can be overcome, the national quota approach may be very useful in many situations.

In summary, resource-related techniques may frequently be important for the optimization of biological yields. But, with the possible exception of the national quota technique, these controls tend to have extremely damaging economic consequences and are of dubious value for the enhancement of employment opportunities.

Regulations Governing the Kind of Fishing Effort

As in the case of resource-related controls, the desirability of regulations governing the kind of fishing effort depends upon the purposes for which they are adopted. In some cases, the problems of incidental catches can be alleviated by the use of fishing gear that is more highly selective than that presently in use. In other cases, it may be desirable to avoid taking fish below a certain size or fish that have not yet reached the reproductive stage. In these instances, the optimum biological yield may require restrictions on the kind of fishing gear or techniques that can be used. As an extreme example, dynamite and other explosives should be prohibited because of their lack of selectivity in terms of either species or size.

But, if the purpose of the restrictions is to achieve a certain yield by increasing the costs of catching fish, then controls on kinds of fishing effort are only temporarily effective. This technique has been referred to as the "leaky bucket" approach. If there is a village well with limited flow, there are no problems as long as the water drawn from the well is no greater than the rate of replenishment. But, if the demand for water increases beyond that point, one solution is to punch holes in the bottom of each person's bucket. The greater the demand, the more the holes.

This analogy is illustrative of many controls (called gear restrictions) adopted in national fisheries. A restriction on gear may initially limit the total catch but, as demand for the product increases, more fishermen will enter the fishery and the restrictions will have to become more severe. The economic damages of this technique are obvious,

32

since the restrictions increase the costs of catching fish but do not increase the amount that is caught or the total revenues from the fishery. On the other hand, this technique generally serves to maintain fisheries labor. Because of this, the political pressures to adopt gear restrictions may be quite strong. However, once this path has been chosen, it becomes increasingly difficult to change. The temptation is to respond to all technological innovations by prohibiting them, and to increase the severity of the restrictions as the demand and prices for the product go up. As the imposed inefficiencies become greater, the number of excessive and redundant fishermen also increases, and the more difficult it becomes to adopt economically rational controls.

In summary, regulations on the kind of fishing effort may be desirable where they increase the selectivity of gear, but their adoption for the purposes of reducing or limiting total yields can be extremely costly. Even in cases where they may be used to protect employment opportunities, questions can be raised about the long-run desirability of doing so.

Regulations on the Amount of Fishing Effort

Fishing effort is a function of three factors: (1) the fishing power of the vessels, (2) the number of vessels, and (3) the amount of time spent fishing. All of these factors must be controlled in the aggregate in order to be effective. If, for example, controls are restricted only to the number of vessels and the length of the season, there would be a strong incentive to increase fishing power by the use of larger vessels, bigger nets, or new catching techniques. Similarly, controls only on the kind of gear (fishing power) and length of season would stimulate the use of more vessels. These effects are inevitable as long as the demand for the product continues to increase. This is why, as noted above, the use of gear restrictions and closed seasons is only temporarily effective in limiting the amount of catch.

Controls on the Physical Characteristics of Fishing. There are two techniques for regulating fishing effort: a physical technique limiting the number of "standardized-vessel-days" and an economic technique limiting effort through the use of taxes or user fees. Referring back to Figure 1, the maximum sustainable yield can be obtained by limiting effort to 5,000 standardized-vessel-days. In a simple fishery, under the jurisdiction of a single state and utilized by vessels of similar size and character, it is relatively easy to impose a limit on effort. In the illustration, if it is assumed that each vessel fishes an average of 200 days during the season, the limit could be achieved by licensing only twenty-five vessels. Furthermore, such a limit could be reached in gradual stages, so as to reduce the hardships that would be incurred by eliminating the superfluous vessels. For example, if there are fifty vessels currently operating in the fishery, all could be licensed initially. Then, through natural attrition or through some scheme for buying out license holders, the number could be reduced to whatever is desirable. This approach would permit the achievement of appropriate levels of catch and would improve economic yields, without imposing undue hardships on the fishermen.

In practice, however, where a stock is shared by two or more states and where vessels and gear are dissimilar, there are considerable difficulties in the use of this technique.[2] One of these is the definition of a "standardized-vessel," which requires estimating the relative fishing power of the different kinds of vessels and gear. For example, how many more fish per day can be caught by a 900-ton stern trawler than by a 150-ton side trawler? In order to provide an answer there must be a long history of detailed statistics, sophisticated analysis of the data, and negotiations to reach agreements on the estimated coefficients of each of the various classes of vessels and gear that are involved. It is also necessary to determine the relationships between total amounts of effort and sustainable yields from the stock. And agreement must be reached on the total permissible amount of effort and on the allocation of shares of the total among the states.

An additional, and more difficult, problem emerges from the fact that fisheries are subject to constant change, and that the dynamic forces of supply and demand are likely to disrupt any arrangement that is based on assumptions of static relationships. In the case of this particular technique for controlling effort, the likely result of these forces will be an increase in the fishing power of the vessels in each of the various classes. Thus, if a state is restricted to a certain number of vessels between 150 and 500 tons in size, there will be an incentive to remove the smaller vessels and bring in larger ones. This shift may occur whether or not the smaller vessels are economically more efficient and whether or not the investments in them have been fully amortized.

In response to this development and its depleting effects on the stock, it may become necessary to establish more classes with more narrow definitions in size. But if this occurs, the fishermen will seek other ways to maximize their individual catches—greater horsepower, larger crews, more mechanical equipment, improved fish finding gear, the adoption of techniques to concentrate the fish, etc. Although the

[2]See, for example, Christy, "Northwest Atlantic Fisheries Arrangements: A Test of the Species Approach," *Ocean Development and International Law Journal*, 1973, Vol. 1., Number 1.

rate at which such developments will take place cannot be anticipated, the changes are inevitable as long as the demand for the product persists. And for each change, there will be an attempt to reduce its effectiveness or to prohibit it either because it will mean a diminution in the optimum biological yield or because it will give some fishermen an advantage at the expense of others.

These developments may make it difficult to improve the net economic benefits from the fishery. The incentives to increase fishing power, through substitutions in size of vessel or through changes in gear or technique, will add to the costs of fishing effort. Economically desirable innovations may be prohibited, with the result that costs might be higher than they would be under ideal conditions. Furthermore, the costs of administrating and enforcing highly detailed restrictions on kind as well as amount of fishing effort could be quite high. And to this might be added the costs of increasingly complex negotiations on the changing definitions of fishing power.

These problems have, perhaps, been given more emphasis than they deserve. But the purpose of doing so has been to point out that the technique has limited flexibility and that there may be damaging long-term consequences that are not readily apparent. Such consequences should be given full consideration in the evaluation of alternative techniques.

Controls on the Economic Characteristics of Fishing. The other approach to controls on the amount of fishing effort is to impose taxes, user fees, or some other kind of economic disincentive. Such a technique may appear to be unacceptable at first glance, but it is probably the most effective means for achieving satisfactory economic yields from fisheries over the long run. The use of such controls would not necessarily reduce levels of catch from fully utilized stocks and might even lead to higher yields. Prices to consumers would not necessarily be increased. The fishermen remaining in the industry would receive about the same amount of net revenues as they did before the imposition of the fees, though each would be taking more fish. The states in whose waters the stock swims would be receiving revenues that are currently being dissipated. And the technique offers sufficient flexibility to permit technological innovations and ready adjustments to changes in the factors of supply and demand.

Figure 2 provides an illustration of the way in which economic controls might operate. This assumes the same relationships between effort and yield as shown in Figure 1, together with a hypothetical tax of $500 per ton of fish. This illustrative situation shows that the tax reduces the total revenues relative to amounts of fishing effort, thereby lowering the total revenue curve (TR). It now intersects the

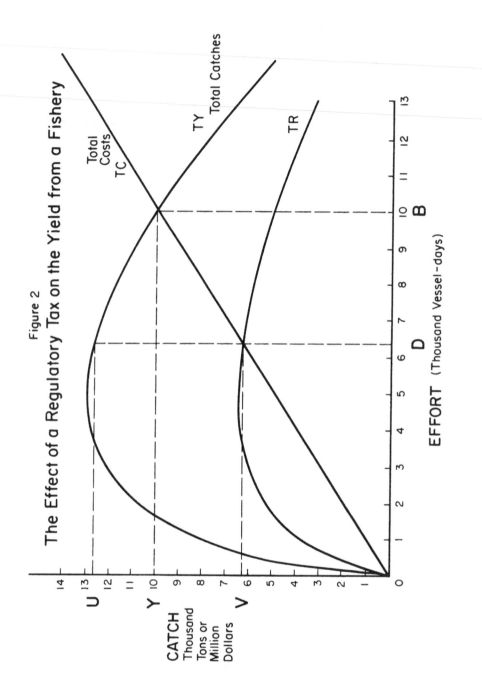

Figure 2

The Effect of a Regulatory Tax on the Yield from a Fishery

total cost curve (TC) at point D of fishing effort. At this point, only about 6,000 vessel-days of effort are employed and total costs of fishing are about $6 million. The amount of the catch has increased from 10,000 to somewhat more than 12,000 tons and the total revenue from the fishery is about $12 million. Of this, about half goes to the fishermen, producing average revenues per vessel-day about the same as they were before the imposition of the tax. The other half would be collected by the management agency and could be used to cover the costs of management as well as some monetary returns to the states sharing the stock. Similar results could be produced through the use of a license fee instead of a tax.

In this illustration, there are a number of simplifying assumptions about uniformity in costs, yields, and prices. The illustration may also overestimate the savings that could be achieved with monetary controls. But the effects of the controls still hold true. In any overutilized fishery, a monetary disincentive will remove some of the superfluous fishing effort; will produce some economic returns to society; and will not leave the fishermen any worse off (or any better off) than they were before. There will be fewer of them, but each will be taking a greater share of the catch than before.

Whether or not prices to the consumers will rise will depend upon a number of factors that are too complicated to consider here, but it is not necessarily true that prices will, in fact, increase. If, for example, the fishermen attempt to pass their taxes on to the consumer by charging higher prices, the amount of consumption is likely to decrease, thereby decreasing total revenues and forcing some of the remaining fishermen out of the fishery. With higher catches per unit of effort, competition among the remaining fishermen may lead to a subsequent price decline. The price at which supply and demand will come into equilibrium cannot be easily anticipated, but it will not necessarily be higher than it was before the imposition of the taxes or user fees.

One of the major advantages in the use of monetary controls lies in their flexibility. The amount of fishing effort and the level of catch can be changed by varying the tax or fee. Innovations can be permitted to take place in response to ordinary economic incentives rather than in response to artificial classes of different kinds of fishing effort. The controls can be established gradually, with small increments in the fees, so as to reduce hardships among the fishermen. In addition, taxes and fees can be used to alleviate some of the problems associated with interrelated fisheries. For example, a high tax might be placed on the catch of incidental species in order to encourage the use of more highly selective gear. Or certain kinds of gear might be taxed at a higher rate than others, in order to reduce conflict.

While the advantages appear to be great, the practical problems of adopting monetary controls for the regulation of fisheries are also great. Perhaps the most important of these is the political difficulty of charging fees or levying taxes for a resource that has traditionally been considered free. This difficulty depends upon the situation, the history of use, and the number and kinds of countries that have been involved. For example, it might be relatively easy to adopt monetary controls where a stock falls within the waters of only two coastal states and there is not a long history of use by distant-water fishermen. On the other hand, in a region such as the Northeast Atlantic, where there are several coastal and distant-water states and where stocks have been fished for centuries, it could be much more difficult to reach the necessary agreements. In the latter case, no individual state has sufficient authority to force the adoption of such measures, and few if any states would be willing to grant sufficient authority to a single management agency. Between these two extremes, there is a wide variety of situations where authority over the use of shared stocks might be acquired and exercised to some degree.

In a traditional fishery, the imposition of taxes or fees is also politically difficult because of the hardships it might place on fishermen whose incomes are already low. Even though the eventual outcome would not be detrimental to the fishermen remaining in the industry, the transitional problems would be very great. There are, however, techniques for alleviating such hardships. Monetary controls can be employed incrementally in combination with direct restrictions on the physical amounts of effort, so that the fishermen do not experience immediate losses.

As in any technique attempting to achieve the goal of optimum economic yield, there would be opposition because of the effect on employment opportunities. Whether or not the sharing states choose to adopt the economic goal depends upon the values they place on employment opportunities. But, if they do choose to optimize economic yields or even to improve economic returns, monetary controls are preferable because they provide revenues that can be used to compensate the fishermen who are excluded or are prevented from entering the fishery.

In addition to the political problems, there are certain technical difficulties associated with the use of monetary controls. If a license fee is used rather than a tax on catch, there are problems in determining what to license. These will be similar to those described above with regard to controls on the physical characteristics of fishing. That is, the fees should be varied in accordance with the fishing power coefficients of the different kinds of vessels and gear. Generally, it would be preferable to adopt a tax on catch, but this raises problems for

38

enforcement and collection of revenues.

Effective use of monetary controls requires a unified approach to the setting of fees or taxes and a high degree of stability over time. If a stock is shared by several states, the fees charged by the different states should be proportionate to the values of catching the fish in the different waters. If they are not, effort will flow to the area where the fees are proportionately the lowest and the optimum economic yield will not be reached. The determination of appropriate fees is a difficult task, not only because of the likely variations in value of fishing the same stock in different areas, but also because of the necessity for agreement among the sharing states or for the acquisition and exercise of sufficient authority by a single management agency. Some of these difficulties could be alleviated by an empirical approach. Fees or taxes could be increased gradually in small increments allowing time to examine the effects and permitting adjustments where these appear desirable.

An alternative technique for determining the appropriate fees would be to auction off licenses to fish or shares in the stock, thereby letting a competitive market determine the relative values of the stock in the different waters. This, however, may create undesirable instability if the rights to fish are not assured over a sufficient length of time to permit adequate returns on capital investments. Fishermen would be unwilling to pay very much for a right that is limited to a year or two unless they can find alternative uses for their vessels in other areas. It is difficult to estimate the length of time or tenure that would permit both the most effective use of capital and the greatest benefit of a competitive market.

These characteristics of monetary controls have been addressed in speculative terms because experience in their use has not been great enough to permit more definitive assessments. The speculations, however, are sufficient to indicate that monetary controls have many distinct advantages over other techniques for the regulation of fisheries. The most important advantages are their ability to achieve optimum economic yields and the flexibility with which the controls can be employed. The most important drawback to the use of monetary controls is the necessity for a high degree of authority on the part of the managing states or management agency.

Combinations of Regulatory Techniques

The discussion of alternative regulatory techniques has been oversimplified by ignoring the possibility of using various combinations of the different methods. In practice, however, most regulations are not purely of one type or another, and the growing complexities of management are likely to lead to greater variations and combinations in the future.

In the Northwest Atlantic, for example, the system involves a melange of closed seasons and areas, total and national quotas, restrictions on gear, and now a proposal for limits on fishing effort; everything except monetary controls. Quite a different pattern is emerging for the management of Canada's salmon fisheries in the Pacific. Although this is essentially an intranational fishery, the developments taking place will provide valuable lessons for the adoption of controls on fishing effort.[3] In addition to the more conventional controls, Canada has imposed a direct limit on the number of vessels, in combination with a system for the extraction of economic rents. It is intended that superfluous effort will be removed on a gradual basis and that license fees will be increased in small increments to reflect the growing value of the license without imposing hardships on the fishermen.

Numerous other variations and combinations of techniques are emerging, and will emerge, in response to the different interests, pressures, and resource characteristics. There are no simple solutions nor solutions that will be uniformly applicable throughout the world. Instead, the arrangements will be marked by considerable complexity, wide diversity, and varying degrees of effectiveness. Although the precise results of any set of regulatory techniques cannot be fully anticipated, careful examination of the alternatives and their likely consequences will help in improving the choices and making the arrangements more effective.

SUMMARY

In examining the alternative regulatory techniques, it becomes clear that they differ quite widely in terms of their effects on the goals of optimum biological yields and optimum economic and social yields. With regard to the goal of optimum biological yields, the total quota (with or without national quotas) is the most effective since it directly limits the catch to the desired level. The other techniques, being indirect, have to be continually adjusted in order to maintain a certain yield. An increase in the price of the product will lead to an increase in the amount of fishing effort because more fishermen will be willing to pay the tax or license fee, or to incur the costs of gear restrictions.

[3] For an excellent account see Peter H. Pearse, "Rationalization of Canada's West Coast Salmon Fishery: An Appraisal," in *Economic Aspects of Fish Production* (Paris: Organisation for Economic Cooperation and Development, 1972), pp. 172–202.

In these cases, effectiveness depends upon the facility with which the regulations can be adjusted.

With regard to economic yields such measures as total quotas, closed seasons and areas, and gear restrictions are not only ineffective but actually quite damaging, because they serve to increase costs without producing any concomitant increase in catches or revenues. The most effective techniques for achieving this goal are those that control the amount of fishing effort. This can be done directly by limits on the number of vessels or indirectly by the use of monetary controls. Here again, there is requirement for flexibility to make continual adjustments to meet changes in the markets for the product. The use of national quotas is only effective to the extent that each shareholder is able and willing to control the effort of its fishermen.

The alternative techniques also have to be examined with regard to the goals of reduced management costs and acceptable patterns of wealth distribution; that is, measured against the costs of research, enforcement, and negotiation. There are no clear-cut measures of such costs, and the estimates can only be expressed in general terms. With regard to research, the question is one of the degree of information that is satisfactory for the adoption and imposition of the regulation. The most costly of the techniques is likely to be the one that limits effort through the use of controls on the physical characteristics of fishing. This requires not only satisfactory estimates of the yield functions of the stocks but also adequate measures of fishing effort and the relative catching power of the different vessels and gear. Monetary controls might be adopted with considerably less information because of their flexibility. They facilitate an empirical approach to management because they can be adjusted in the light of observed results. A total quota can be used in the same way because it, too, permits incremental adjustments.

Enforcement of regulations requires means for surveillance, arrest, and trial. Of these, only the first is directly related to the kind of regulations that are adopted. The cost of surveillance is minimal if it can be done by aerial observation, as in the case of closed seasons and closed areas. Much higher costs are incurred for systems that require inspection of catch and/or gear on board fishing vessels. This might be necessary for licensing schemes and national quotas, particularly in areas where there are high incidental catches.

The costs and difficulties of negotiation depend upon a variety of factors in addition to the choice of regulatory technique. Negotiation is generally easier where there are fewer states, where the states have similar perceptions of the need for the controls, and where the states have common goals and interests. But with any given set of conditions, the costs of negotiation will depend upon the kind of technique that is adopted. Controls that do not directly affect the distribution of wealth, such as closed seasons and areas, may be easier to negotiate than those that do. Negotiation may also be easier with the use of economic controls that extract the economic rent because they separate the problems of distribution from those of management. The initial adoption of economic controls is likely to be difficult to negotiate, but once in effect, negotiation on the distribution of the rents can be removed from the tasks of management, and distribution can be facilitated by the use of money as a common denominator. Where a common denominator does not exist and where the regulation directly affects distribution, negotiating costs are likely to be very high. In the Northwest Atlantic, for example, if the U.S. proposal on effort controls is adopted, it will require annual negotiations on (a) the total quotas, (b) the allocation of national quotas, (c) the total amount of fishing effort, (d) definition of fishing power for the different vessels and gear, and (e) national allocations of shares of effort.

In summary, the choice among alternative regulatory techniques will depend not only upon their effectiveness in improving economic and social benefits, but also upon the associated costs of research, enforcement, and negotiation. In addition, choices will have to be made in the light of different management situations which vary widely both with regard to the characteristics of the resources and with regard to the nature of national interests. The diversity of situations and techniques means that there are no uniform solutions to the problems of management.

Although this discussion has emphasized the diversity and complexity of management problems, it has also indicated certain general principles that can be used to facilitate solutions. One of these is the principle of flexibility; the other, the principle of economic authority. Where both principles can be adopted, the tasks of management are greatly reduced.

Flexibility is critical because the conditions of supply and demand are constantly changing and constantly affecting the kind and intensity of fishing effort. And since many of these changes cannot be anticipated clearly, it is desirable to adopt techniques that can accommodate the changes without too much difficulty. In this regard, monetary controls are preferable because they can be easily adjusted in response to changes in the intensity of fishing effort and because they can accommodate new techniques and new entrants into the fishery more readily than other techniques. But even if monetary controls cannot be adopted, it is still important to build in flexibility wherever possible. For example, a system of national quotas should allow for transferability of the quotas in order to permit states to take advantage of economies of scale or permit them to sell shares to new entrants.

42

A tax on the quotas should also be imposed. The amount might be negligible to begin with, but it would establish the principle of extracting economic revenues and permit its eventual use as a means for control.

In addition to flexibility, the principle of economic authority is also important in seeking better solutions to the problems of management. The authority might be vested in the hands of a single coastal state, an agency created by several coastal states, an agency created by a group of coastal states and distant-water states, or possibly an international body. But whatever the case, the greater the degree of authority, the easier it will be to manage the resource. This means that states sharing a resource should adopt techniques that permit, rather than preclude, a gradual increase in the discretionary powers of the manager.

III Alternative Techniques for Distribution of Wealth

Introduction

IN THE PAST, most of the seas' wealth was generally distributed on a pattern reflecting the relative amounts of fishing effort of the individual states. Limits of jurisdiction were narrow and, for the most part, access to fisheries beyond the limits was free and open, meaning that only those states which exercised their rights to fish received the wealth. But with the growing demand for fish and the increasing value of fishery rights, the patterns of wealth distribution are being shaped by forces other than investments in fishing effort. This is occuring not only through extensions in the limits of jurisdiction but also through various kinds of bilateral and multilateral agreements which tend to be exclusive in nature, such as abstention arrangements and national quotas. The exclusive arrangements deal explicitly with the problems of distribution. Extensions of jurisdiction frequently lead to the same result, because neighboring states are forced to reach agreements on the shared stocks that fall jointly within the limits of their claimed areas. Thus, it is becoming apparent that the most significant fisheries problem for the next decade will be that of resolving conflicts over "who gets what?" from the seas' wealth in fisheries.

The problems of distribution are particularly difficult to deal with because of the lack of clarity and uniformity about the goals and criteria for distribution, the definition of wealth, and the interests of the participants. Such problems do not lend themselves readily to critical analysis, for there are no rational, objective criteria for determining who should get what shares of wealth. The most that can be said is that distribution should be acceptable; that is, that all relevant states are better off by agreeing to the arrangement than by violating it. It is to be hoped that the patterns of distribution will also be equitable and meet some general sense of fairness. And it is desirable that, in achieving acceptability, there should not be a significant sacrifice in the production of net values.

But it is difficult to find useful guides for determining what is, or is not, acceptable. This is essentially a matter of negotiation, and one that will involve many other values than those directly related to fisheries. However, it may be useful to examine some of the criteria that have been suggested for distribution and to classify the different interests of different states. These elements are discussed below.

Criteria for Distribution

As stated, there are no rational, objective criteria for determining patterns of distribution. Many criteria have been suggested but they tend to be subjective, imprecise, and provide poor guides for the making of decisions. Some of the ones most frequently advanced are noted below:

Proximity. The proximity of a stock to the coast of a state is generally the most important criterion suggested for claiming a share or "preferential right" in the seas' wealth. Within very narrow limits, this criterion obviously carries considerable weight. But there is no generally acceptable definition of the limits and no rationale for choosing 3, 12, or 200 miles. Furthermore, proximity does not resolve problems of distribution in those situations where stocks migrate along coastlines through the waters of neighboring states.

Manageability. It is sometimes stated that because a state is in a better position to manage a stock than other states, it should acquire not only exclusive rights to manage but also exclusive rights to the wealth of the stock. There are several situations in which such arguments might be advanced. With regard to demersal (bottom feeding) species, the depth of the continental shelf or slope might provide a natural limit to the outward extent of a species' movement. For anadromous species (those spawning in fresh waters), the manageability criterion might apply several thousands of miles from shore, wherever the stock migrates. And in terms of certain pelagic (surface feeding) species, it has been argued that they inhabit a coastal environment that is indivisibly related to the terrestrial environment and therefore a form of "natural prolongation" of the land that justifies exclusive jurisdiction.

It is clear that management of a stock, wherever it occurs, should be subject to unified control. In certain situations, it is likely that the control can best be exercised by a single state. But it does not necessarily follow that the ability to manage a stock need be accompanied by allocation of the entire proceeds of exploitation. It is quite possible to adopt arrangements that leave management in the hands of a single state while allowing for a sharing of benefits among several states.

Historic Rights. Past patterns of use are often advanced as a basis for future patterns of distribution. But historic rights are difficult to use as a criterion for distribution, because there is no uniformly acceptable definition of history. In the case of the Northwest Atlantic, for example, national quotas have been allocated partly on the basis of ten years of experience and partly on the basis of three years of experience. While this definition of history was acceptable to the states involved in the decisions, it has no necessary relevance elsewhere, and it might just as well have been twelve and two years, eight and four, or any other combination.

Need. Many states have argued that they should acquire a larger share of wealth because they have a greater need or dependence upon the stock for food, employment opportunities, income, reduction of import requirements, etc. But while there may be certain humanitarian values in distributing wealth according to need, the application of this criterion is extremely difficult. What percentage of income, protein, or employment should a state derive from a resource in order to qualify as being dependent upon it? To which resources should the criterion apply? If need is accepted as a criterion, why should it not apply to resources far distant from its shores as well as to resources that are close?

Capacity to Exploit. An example of this criterion can be found in the U.S. Draft Fisheries Article where it states that "the coastal State may annually reserve to its flag vessels . . . that portion of such coastal and anadromous resources as they can harvest." This criterion gives the appearance of a greater degree of precision than the others, because quantity of catch can be measured, more or less, in terms of quantity of effort. But it requires an adequate definition of what constitutes a coastal species—a definition that is particularly difficult in such an area as the North Sea or the West African coast where the same stock can be found in the waters of different states at different times in its life cycle. Furthermore, it provides no basis for distribution when the combined efforts of the states in whose waters the stock swims are greater than that which is appropriate to the maximum sustainable yield of the stock. In addition to its inadequacy as a criterion for distribution, the principle that it would establish may be extremely damaging to the production of economic benefits from the resource. The criterion of exploitability provides an incentive for coastal states to expand their fishing effort, in order to acquire a greater share of the stock for themselves. The incentive would exist as long as the state were receiving less than the total allowable yield from the stock. But, as noted above, fishing effort is already excessive in many areas

46

throughout the world and the encouragement of greater amounts of coastal state effort (even though this may be matched by declines in foreign effort) is not conducive to reaching the goal of optimum economic yield.

Common Heritage. A final criterion for distribution (although one not yet applied) is that the resources of the sea are the "common heritage" of mankind and should therefore be shared by all states of the world. This provides no guidelines for the determination of the amount of shares in wealth, but it does assert that those states that do not have satisfactory access to fisheries should, nevertheless, receive some of the values that are produced. It should be noted that, unlike the other criteria, this one can only be met if some of the economic values of fisheries are extracted from the users. For example, distributing a right to fish or a national quota to a land-locked state is of no value to that state unless it can extract some benefit by selling or leasing its right.

Summary. Other criteria and variations of the above are likely to be brought forth from time to time. In any particular situation, certain of the criteria may be adopted; that is, they may produce patterns of distribution that are acceptable to the relevant states. In the Northwest Atlantic, for example, the fifteen member states of the International Commission agreed to distribute the wealth of fourteen stocks of groundfish on the basis of preferential rights of coastal states and historic rights of distant-water states. The distribution was based on an arbitrary formula of "40-40-10-10." Forty percent of the total allowable yield of each stock was distributed on the basis of the average catches of the states during the past ten years; an additional 40 percent was distributed on the basis of catches during the past three years; 10 percent of the yield was reserved for the coastal state (the United States or Canada, depending upon the location of the stock); and a final 10 percent was distributed on the basis of special needs or to allow for the interests of nonmembers. There is an apparent symmetry to the formula of "40-40-10-10," but there is no basis for saying that it is a better or a worse pattern of distribution than any other formula.

As this example illustrates, criteria are negotiable, and their importance lies not so much in their value as guides for decision as it does in their implications for resulting patterns of distribution.

Four Sets of Interests

For any particular stock of fish, four different sets of interests might wish to claim some share in the benefits from the stock. First, of course, there are the interests of the state (or states) that are most nearly adjacent to the stock. If the stock does not fall entirely within their jurisdiction, they may still wish to claim a preferential right to a share of the stock. This may mean a right to fish for a guaranteed share of the yield or a right to receive a certain share of revenues from the users.

The second set of interests includes those states which have had a history of exploiting the stock. These states feel that they have an historic right to the resource and that they should not be excluded from continued enjoyment of that right. These states are referred to here as distant-water states, although in many cases the distances may not be very great.

A third set of interests can be found in the newly emerging fishing states, those that are beginning to develop distant-water efforts and that do not wish to have their opportunities for fishing diminished. These interests—referred to here as "third-party" interests—are likely to oppose exclusive arrangements, whether these are unilateral extensions of jurisdiction or national quota agreements that do not provide for new entry. States with these interests include the ones that are experiencing rapid rates of increases in catch but that do not have particularly large resources off their own coasts.

The fourth set of interests includes all the other states—those that are not close to the stock and that neither fished for it nor intend to do so in the future. Such states may wish to claim a "common-heritage" interest in the resources, receiving some form of benefit from the use of the stock.

These sets of interests have been expressed with regard to a particular stock of fish. It is clear, however, that some states will hold several, or perhaps all four, of the interests if all stocks of fish are included. That is, a single state may wish to claim a preferential right to a stock close to its shores, an historic right in fishery off foreign shores, a third-party interest in another distant-water stock, and a common-heritage interest in stocks much further afield. While such disparate interests do exist within certain states, many if not most states are likely to find that their predominant interest lies in one or the other of the four sets.

Looked at from the point of view of the stocks, it can also be seen that different situations will elicit different degrees of interests. Stocks that occur only within the internal waters of a single state will clearly be of interest only to that state. On the other hand, stocks (such as Antarctic whales) that do occur far out on the high seas may be of little interest to coastal states and of strong interest to states claiming rights of common heritage.

Enumerating these four sets of interests, however, does not mean that all of them must necessarily be accommodated. Indeed, no suggestions can be offered as to whether or not they *should* be accommodated, because there are no adequate criteria for making such a suggestion. It is simply pointed out that these various interests exist and that it may or may not be necessary to accommodate them to varying degrees in order to achieve acceptance of arrangements.

It should be noted that this discussion of the different sets of interests can apply to a region as well as to the entire ocean. Thus, landlocked states might invoke a common-heritage interest in the resources of a nearby sea but not for the world as a whole. Or the third-party states bordering that sea might be given certain rights or shares of wealth that they would not receive in other oceans.

WEALTH DISTRIBUTION TECHNIQUES

A wide variety of techniques can be used for the distribution of the seas' wealth in fisheries. These techniques, however, can be divided into two major categories: those that distribute the resources and those that distribute the benefits from the resources. The first category defines a share of wealth as a share in access to the resource. Wealth is acquired only when fishing takes place. The second category includes techniques which permit a state to receive some benefits from a resource without having to fish. The distinction is important because of the implications for accommodating the various sets of interests and because of the significance for the choice of alternative arrangements. The two techniques can, of course, be combined.

Within the first category—distribution of resources—the two major approaches are the zonal and the species techniques. These are often posed as the two major alternatives. But the differences between them are far from clear and appear to lie more in the kind and degree of authority that can be exercised by coastal states than in the means for the distribution of wealth. With regard to the second category—distribution of benefits—there is little real difference between the various techniques. The benefits can be expressed as economic values, derived from the sale of quotas or rights to fish or derived from the use of taxes or license fees. They can also be expressed through the use of joint ventures or forms of economic aid such as port developments, training of fishermen, etc. Distribution of products, except in the unique case of the North Pacific fur seals, is generally infeasible.

Resource Distribution by Zones

Significant changes in the patterns of distribution of wealth have been accomplished in recent years by the zonal approach. Within the past decade, the number of states claiming limits of jurisdiction greater than 12 miles has grown from about 8 to more than 29. The amount of wealth gained by the use of zones, however, is only partly related to the extent of the claim. It depends also upon the value of the fishery stocks in the waters, the degree to which they are shared with neighboring states, the ways in which the jurisdiction is exercised, and the costs associated with the exercise of jurisdiction.

A zone based on distance alone has no economic or biological rationale, because the habitats of fish species do not conform to arbitrary, man-made boundaries. Zones, however, can also be based on depth, and here there is at least a partial biological justification with regard to certain stocks of sedentary and demersal species, most of which do not exist beyond a limit of depth somewhere along the continental slope. For many pelagic species, however, neither depth nor distance has any particular relevance.

There is considerable variability in the value of zones because of the wide differences in natural fertility and in length of coastline. A limit of 12 miles in one area of the ocean, for example, may have much greater worth than a limit of 200 miles in another area. There is, thus, no uniform limit that would provide an "equitable" distribution of wealth among all coastal states.

The zonal approach provides for exclusive jurisdiction over *all* marine organisms within the zone, whether these resources are utilized fully, partly, or not at all. It does not, however, necessarily provide for unilateral jurisdiction over single stocks, because of the migrations of stocks into the waters claimed by other coastal states. In these situations, the wealth that is acquired by the use of the zone must still be shared with others, and other techniques for distribution must be used. Since this is a fairly common situation in many areas of the world, the zonal approach is only partially effective in distributing wealth.

In addition, certain species may extend beyond the limits of the zones into international waters, depending upon the extent of the limits and the migratory characteristics of the species. Certain species of tuna, swordfish, whales, seals, and anadromous species such as salmon may fall into this category, requiring the adoption of other techniques for distribution in addition to the zonal approach.

The amount of wealth acquired under the zonal approach is also significantly affected by the way in which jurisdiction is exercised. The zone does not necessarily (or even generally) mean the exclusion of foreign fishermen. In some cases, foreign fishing could continue to be free and open for stocks not utilized by the coastal state. Or, the zonal approach might be combined with other techniques; national quotas, for example, might be adopted within a zone, permitting the distant-water fishermen to continue to receive a share of the wealth. Provisions might be made for phasing out the distant-water fishermen as the capacity of the coastal state fleet increases. Economic measures might also be employed, so that the distant-water fishermen could maintain access subject to the payment of a license fee or royalty or subject to their investment in joint ventures. In short, there is a wide variety of different ways in which coastal states can exercise jurisdiction within a zone and allow for different patterns of wealth distribution.

In general, however, the zonal approach tends to shift patterns of wealth distribution in favor of coastal states at the expense of distant-water interests, and it makes no specific allowance for the accommodation of third-party or common-heritage interests. Although coastal states are likely to receive the greatest benefits from a zonal approach, there is a wide disparity in the amounts that each state might gain. To those states fortunate enough to have both long coastlines and rich offshore fisheries, the gains might be very large. Several North and South American states fall into this category. But there are many other coastal states that do not have both attributes and, in addition, many states that are "shelf-locked"; that is, with limited vistas because they are on small seas or faced by foreign islands. Whether or not the coastal states as a whole are willing to adopt a general rule on the zonal approach depends upon how they perceive their own net gains and also upon how they perceive their gains relative to others.

The effect of the zonal approach on distant-water interests is generally negative, but the loss to distant-water fishermen is by no means directly correlated to the extent of the zone. As indicated, in some cases they might continue to enjoy free and open access. But even if they are required to pay for the use of the fisheries (through license fee, royalty, lease fee, or some form of investment in coastal state operations), the net losses may not be commensurate with the costs of the fees. In the case where the stock has been depleted, the extraction of economic rents within the zone may actually lead to higher total catches. The fishermen willing to pay the fees will be no worse off than before because there will be fewer of them and catches per unit will be higher. The distant-water states may, however, experience some costs in making adjustments for the fishermen who are displaced by the imposition of the fees.

If the stock is not being utilized up to the point of maximum sustainable yield, the fees may lead to lower total catches, depending upon the degree to which they affect the amount of fishing effort. In some cases, the fees may simply extract some of the surplus profit; in other cases, they may make fishing uneconomical for some of the fishermen, thereby reducing total effort and leading to lower total catches.

It is quite difficult to anticipate the net effect, not only because of the inadequacy of economic information but also because of the different degrees to which states respond to economic forces. It is useful to emphasize, once again, the fact that fishing industries are notoriously subject to other forces than those of the market. Because of this, some may refuse to pay any fees while others may be willing to incur costs that are not economically justifiable.

The zonal approach provides no specific means for the accommodation of third-party interests. In those situations where wealth is defined solely in terms of access or a right to fish, the opportunities for meeting third-party interests may be severely limited. Preferences would probably be given to states with historic rights, while third parties might only be allowed access to stocks not fully utilized. If, however, wealth is defined in terms of the benefits from fishing, third-party interests would be on the same footing as those of present distant-water fishermen. Defining wealth in terms of benefits permits the use of economic measures to determine access. Such measures make no necessary distinction between states, and those willing to pay the most would gain the right to fish.

Common-heritage interests cannot possibly be met by a zonal approach where wealth is distributed solely on the basis of access since, by definition, common-heritage states are those that do not wish to fish for the particular stock. The wealth must be extracted in some form that is shareable, such as economic revenues, in order to permit distribution to common-heritage interests. This does not mean that common-heritage interests must, necessarily, be accommodated. It simply means that, if they are, then the zonal approach must be combined with an economic one.

Resource Distribution by Species

The proposals by the United States and others for a "species approach" are not directly, in themselves, techniques for distribution. They are aimed primarily at the problems of management, recognizing the different characteristics of different kinds of fish and the necessity for unified controls over individual stocks wherever those stocks are found. But under the proposals, it is clearly intended that distribution of wealth be based on the resources as a whole, rather than on geographic zones. The primary technique for distribution, under these proposals is apparently that of national quotas—the distribution of the yields of particular stocks among the participants. There are variations of this technique, ranging from an implicit distribution based on present patterns of catch to abstention, which allocates the entire quota to a single state.

Under the species approach proposal, three kinds of fish are distinguished: anadromous, highly migratory oceanic, and coastal. It is suggested that there be separate treatment for the management of these groups, but little is said explicitly about the problems of distribution. With regard to the anadromous species, it is pointed out that the host state should be assured of a sufficient catch (or return) to justify its management costs. This leaves open the question as to whether the host state should acquire the total yield from the stock, with other states abstaining, or something less than the total. If it is less than the total, it also leaves open the question as to how the balance should be distributed. With regard to the highly migratory oceanic species, no guidelines have, thus far, been offered for distribution. Presumably, distribution would be through national quotas based primarily on past and present histories of catch. If the current developments in the Inter-American Tropical Tuna Commission are adopted as precedents, some preferential rights may be accorded to nearby states.

The most important group is that of the coastal species, and here various explicit suggestions about distribution have been made. In essence, these suggestions are for gradual changes in the status quo, and for distribution on the basis of negotiation rather than on zones. The coastal states are granted a preferential right to the stocks off their shores—a right to be based on their "capacity" to take the fish. The distant-water states are not to be excluded but can continue to acquire shares of the wealth in the form of quotas. If a coastal state is not utilizing a stock off its shores, it cannot exclude foreigners from access. If the coastal state is utilizing the stock, access by foreign fishermen may be phased out as the capacity of the coastal state increases. Other countries have suggested that there should be limits to the phasing out of distant-water fishermen, depending upon whether the coastal state is developed or developing, or depending upon the degree to which the local fishermen are dependent upon the stock. It might be pointed out that the introduction of such criteria inordinately complicates the process of negotiation.

But whatever the criteria, distribution is achieved primarily through the allocation of access to the resources. A national quota means that a state can catch a certain quantity of fish and that, when it reaches that limit, it must stop fishing for that stock until the next season. Presumably, it can use that quota in whatever manner it wishes, investing a large amount of effort for a short period, or a small amount of effort over a long period. It is theoretically free to prohibit technological innovations and maintain employment opportunities, or to optimize its economic yield by limiting effort to the most efficient vessels.

A variation on national quotas of catch is that of national allocations of fishing effort. Under this technique, a total limit is imposed on the amount of fishing effort that can be employed, and shares of the total are distributed among the participating states. In a section of the East China Sea, for example, Japan and China agreed in 1970 to limits on the number of seiners that each could use.[1] And, as noted above, the United States has submitted a proposal for effort limits in the Northwest Atlantic. The difficulties of this approach have already been discussed.

Distribution of wealth through the technique of national quotas on catch or national allocations of effort can be combined with other distribution techniques. One of the easiest methods is that of permitting transferability of the shares, so that a state can sell or lease its entire share or a portion of it. This permits a state to extract an economic benefit from the right it acquired in the initial distribution of shares. In the case of the Antarctic whales, for example, national quotas were associated with the numbers of fleets of the different states. Transfer of fleets and quotas was permitted and Japan, thereby, was able to increase its share of the yields by purchasing the fleets of the United Kingdom and the Netherlands. The purchase price presumably reflected the values of the resource rights to the various parties.

If national quota schemes are used, transferability of shares is particularly important. Without this, there is a considerable loss in flexibility. Adjustments in intensity or scale of effort in response to changes in the factors of demand and supply are severely impeded. States tend to become locked in, because they receive nothing in return for decreasing their shares, and can only increase shares by negotiation. The system does not accommodate changes in comparative advantages among states. In addition, prohibitions against transfer increase the difficulty of resolving conflicts in the use of interrelated species. For example, where there are problems of incidental catches, the state whose interests are damaged can do little but attempt to renegotiate the quotas. If quotas were transferable, however, accommodation of the different interests could be facilitated.

Another technique for distribution is to combine catch quotas with taxes, or effort allocations with license fees. This would permit much greater flexibility in the distribution of wealth. However, this can only

[1]Choon-Ho Park, "Fisheries in the Yellow Sea and the East China Sea," unpublished ms., March, 1973, p. 42.

be achieved where an individual state, or, possibly, an agent acting for a group of states, has a high degree of authority. In the zonal approach, it is not difficult to imagine an individual state exercising such authority. In fact, this is already being done in several instances. But it is questionable whether the proposals for the species approach, by the United States and others, intend to provide coastal states with sufficient authority to impose taxes or user fees. Although the U.S. proposal suggests the possibility of extracting revenues from foreigners to help the coastal state defray the expenses of management, it is apparent that distribution of coastal species is to be achieved primarily, if not entirely, through the use of national quotas or, possibly, national shares of effort.

With regard to the four major groups of interests, the species approach appears to have different distributional effects, initially at least, than the zonal approach; more closely approximating the status quo. Coastal state shares of wealth are restricted to preferential rights based on their capacity to fish. They cannot, as in the case of the zonal approach, readily extract economic rents from foreign users and achieve shares of wealth in terms of benefits. The distribution of wealth to distant-water interests is not significantly curtailed since they can continue to fish for the shares not fully utilized by the coastal states without being required to pay for the privilege.

New distant-water states with third-party interests cannot readily be accommodated through the species approach, except in cases where the stock is not being fully utilized. In that situation, they presumably would have free and open access and could achieve a share of the wealth without having to pay for it. In the case of full utilization, however, where total yields are already divided into national quotas, distribution to third parties would be severely limited. In any particular situation, a portion of the yield could be reserved for new entrants, but it would be difficult to determine how much to reserve and what to do if and when that portion becomes committed. If there is no reserve or the reserve becomes allocated, additional third parties could only be accommodated by diminutions in the shares of present members.

Entry by third parties could be facilitated by permitting the transfer of quotas, so that a new state could buy its way into a fishery. This, however, forces the new state to pay for something that the other states have received free because of their proximity to the resource or their historic rights. As a general principle, this would mean essentially a distribution of wealth based on past use and historic rights.

The species approach, like the zonal approach, makes no provision for meeting the common-heritage interests and actually precludes this possibility as long as wealth is defined solely in terms of access.

In general, both approaches tend to concentrate primarily on the distribution problems between coastal and distant-water states, allowing for little or no accommodation of third-party and common-heritage interests. The zonal approach tends to distribute relatively greater shares of fisheries wealth to the coastal state, while the species approach tends to give more protection, initially at least, to the distant-water interests.

A more significant difference between the two lies in the kind and degree of authority that can be exercised over the use of the coastal species. The extension of jurisdiction provides coastal states with full rights to dispose of the resources as they see fit. They have rights not only to control the use of the resources within their zones but also to appropriate the wealth either by excluding foreign fishermen or by extracting economic revenues and benefits. Under the species approach, the coastal states have a high degree of authority, but it is restricted to the imposition of physical regulations and controls and, as currently envisaged, does not permit them either to extract economic revenues or determine the distribution of wealth.

The restricted authority of coastal states under the species approach means that negotiating costs are likely to be much greater. First, it will require complex negotiations over the distribution of wealth; i.e., the determination of national quotas or national shares of effort among several parties. Similar negotiations will be required in the zonal approach where stocks migrate between the waters of two or more coastal states. But in other situations, negotiations by coastal states with extensive zones of jurisdiction will be minimal and need not be much more difficult than those over the use of terrestrial resources.

Second, negotiating costs under the species approach will be much higher because of the restricted definition of wealth. Where wealth is measured solely in physical terms, such as quotas or access, it means that distribution is essentially a barter of items that may be valued quite differently by the different participants. This is particularly difficult if it is necessary to define the meaning of fishing effort and the fishing power of disparate kinds of vessels and gear. It is also particularly difficult where stocks are interrelated. Furthermore, negotiations will be greatly complicated by changes in the factors of supply and demand, in comparative advantages among states, and in national capacities. These changes will necessitate frequent renegotiations.

The difficulties of using economic controls under the species approach, as it is now proposed, will also limit the choice of management techniques and reduce flexibility. The management techniques will be restricted to those that preserve some degree of access to the participants—total and national quotas, closed seasons and areas, and negotiated physical controls on the kind and amount of fishing effort. With the exception of catch quotas, these techniques tend to become increasingly inflexible and more restrictive as demand for the products increases.

Distribution of Resource Benefits

Instead of sharing in the resource itself, it is possible to share in the benefits from the resource. This technique, as noted above, can be used in combination with either the zonal or the species approach, provided that there is a single state or a single unified agency that has the authority to make the users pay for the use of the resource. Payment is most easily expressed in some monetary form, through the use of taxes, license fees, or leases. But it might also be made indirectly by requiring the user to invest in joint ventures or contribute in some way to the local economy. It is even possible, as in the case of the fur seal agreement in the North Pacific, to express payment in terms of the product that is harvested.

The choice of which of these various benefits should be extracted from the users depends upon how the benefits are to be distributed. Where a stock lies fully within the zone of a single coastal state, that state can use any of the different techniques. But where a stock is shared, the benefits to be extracted must also be shareable. This would generally mean the use of taxes or license fees, since it is far easier to distribute money than to share in port development, technological aid, or the actual products of the fishery.

The use of this technique for distributing wealth depends upon two factors: the desirability or necessity for doing so in order to arrive at an acceptable arrangement and the competence or authority for extracting benefits from the resource users. These two factors are discussed below in relation to the three kinds of species—anadromous, oceanic, and coastal.

For anadromous species, such as salmon, the question of distribution is raised by their appearance on the high seas far from the waters of the host state and by the history of use that has permitted other states than the host state to participate in exploitation. There are current pressures by the host states to change this pattern of distribution and enforce the principle of abstention, thereby acquiring the total benefits of the resource. If these pressures are successful, either through the adoption of a general principle at the UN Conference or by regional agreements, then the question of distribution is resolved. But the pressures may not be successful, and the nonhost states may insist upon receiving some share of the benefits from the resource.

In this situation, it will be necessary to determine how the benefits are to be shared. Agreements might be reached, as they have in the past, providing the other states with national quotas, permitting them to take a certain quantity of fish in the waters beyond the jurisdiction of the host state. This, however, is of questionable desirability because of difficulties it creates for optimizing biological yields from the stocks. For purposes of management, it is generally preferable to take the fish close to the mouths of the spawning streams. Thus, instead of sharing in the resource, it may be desirable to extract some benefits from the resource and distribute these to the relevant states.

The most direct means for doing this would be for the host state to make payments, proportionate to the value of the catch, to the other states that abstain from taking the salmon on the high seas. A less direct, but perhaps politically more realistic, means would be to permit the foreign state (or its companies) to invest in the local industry. Other ways for distributing benefits from anadromous species might also be adopted.

The choice of the technique will depend also upon who is to share in the benefits. If the principle is adopted that a nonhost state has a right to share in the benefits from anadromous fisheries, questions are raised as to whether this should be restricted to states with a demonstrated history of use, or should include third-party states with a potential interest in the stock, or even nonfishing states with a common-heritage interest. In the last case, the only technique is for the host state to provide some form of economic revenues to a generalized fund.

With regard to highly migratory oceanic species, such as tuna, the present techniques for distribution include both a sharing in the resource itself and a sharing in the benefits, though neither of these techniques has been adopted in a formalized manner. In the eastern tropical Pacific, for example, distant-water states receive a share of the tuna by fishing for the stocks. The coastal states receive shares in the form of quotas, but also by extracting license fees or fines from some of the foreign users when they fish within the 200-mile limits claimed by the coastal states.

The movement of tuna stocks in and outside of a zone of 200 miles, together with the desirability of some degree of global control, raises difficult questions for determining what states will share in distribution and how the distribution is to be accomplished. It is conceivable that distant-water states could do all of their fishing for tuna outside of a 200-mile zone, distributing the wealth among themselves in the form of national quotas and making no payments to coastal states. In this event, the coastal states would acquire wealth in the form of whatever fish they could catch within their jurisdiction plus whatever quotas they might receive for tuna outside their jurisdiction.

It is more likely, however, that the present patterns of distribution will be elaborated and that coastal states will receive some benefits from the tuna fisheries whether or not they fish themselves. For this to be effective, there must be close collaboration between and among both the coastal states and the distant-water fishing states. Misallocation of effort, and a loss of potential net benefits, could easily occur if license fees or taxes are not properly coordinated. This suggests the desirability for a regional agency with sufficient authority to set the appropriate fees for the use of the resources. The revenues collected could then be distributed in whatever manner is acceptable. This technique for distribution of benefits would facilitate the accommodation of third-party and common-heritage interests, should this be desirable or necessary.

For coastal species, there are certain situations in which distribution of benefits from a resource may also be preferable to distribution of the resource itself. For example, where a stock swims within the waters of two or more coastal states but is used only by foreign fishermen, distribution would be facilitated by extracting economic revenues from the users. As in the case of tuna, the fees should be coordinated in order to prevent misallocation of effort and a loss of potential benefits. On the other hand, in certain situations, a state may prefer to distribute the resource itself. This might occur where the states in whose waters the stock swims, use or wish to use the resource themselves and where there are no distant-water, third-party, or common-heritage interests that need to be accommodated. In this situation, national quotas might be the easiest and most satisfactory technique for distribution.

In between these two extreme situations, the problems of distributing wealth in coastal species become much more complicated, since both coastal states and foreign states have an interest in using the same stock. To distribute wealth solely on the basis of economic revenues would necessitate the imposition of taxes or license fees on coastal state fishermen as well as on foreigners. While this might be desirable from the point of view of economic efficiency—permitting maximum production of net economic revenues—it would be politically difficult for a coastal state to tax its own fishermen and share these proceeds with other states. Conversely, distribution solely on the basis of national quotas or access to the resource, among both coastal and distant-water states, would be accompanied by difficult problems of negotiation and a possible diminution of net benefits, as discussed above. In addition, it would not readily permit accommodation of either third-party or common-heritage interests.

In this situation, some combination of techniques is likely to emerge, with wealth being distributed partly on the basis of quotas and partly on the basis of economic benefits. The conditions in the different regions are so disparate that one cannot anticipate how benefits are likely to be produced and extracted or to whom they will be distributed.

SUMMARY AND CONCLUSIONS

There is no way to determine which technique or combination of techniques for distributing wealth in fisheries will best meet the goal of acceptability. The techniques provide the means for distribution but do not determine the patterns, although certain approaches may better facilitate the accommodation of different interests than others. Thus, it can be pointed out that common-heritage interests (either regional or international) can only be met through techniques that extract the benefits from the resources. But in this paper it cannot be determined whether, or to what degree, common-heritage interests should be met. Similarly, it can be pointed out that the species approach and the use of national quotas for distribution tend to serve the interests of the distant-water states better than the zonal approach (although this need not necessarily be the case). But, again, there are no criteria for determining the degree to which distant-water interests should be met.

However, even though it is beyond the scope of this paper to determine acceptability, certain general conclusions can be expressed which may be of value in the examination, by states, of the alternative distribution techniques. One conclusion is that flexibility in the techniques is highly desirable. The economic forces affecting fisheries are constantly changing, with the result that new interests are emerging and changes in comparative advantages continually taking place. This means that states at one point may receive their greatest benefits by fishing, but at another time may be better off by sharing in revenues. They cannot make such changes, however, as long as sharing in wealth is inextricably tied to the exercise of the right to fish.

A second general conclusion is that it is highly desirable to reduce the costs and difficulties of negotiations over wealth distribution. In those situations where wealth is defined in terms of access, negotiating costs can be extremely high, because the values of fishing rights are obscured and there are no direct means for compensating those whose access is diminished. High negotiating costs also reduce flexibility, because desirable changes in intensity and kind of fishing effort depend upon agreements rather than ordinary economic forces.

Third, the chief means for providing flexibility and reducing negotiating costs is by the separation of rights of access from rights to share in the wealth, and by the distribution of benefits rather than physical quantities of the resource. If fishing rights are available upon the payment of fees, new states can enter without difficulty, and changes in intensity and kind of fishing effort can be made relatively easily. Negotiations over the distribution of wealth are facilitated, because they can be separated from the decisions on management and because a common value is being shared.

Finally, it can be said that the adoption of techniques for extracting and distributing benefits can only be achieved where a high degree of authority can be exercised. It is this conclusion, however, that is the most difficult to adopt and that is likely to be the major source of controversy at the Law of the Sea Conference. The acquisiton and exercise of satisfactory authority will be extremely difficult to achieve in many situations. The difficulties may, indeed, be greater than the benefits that can be achieved. However, attempts should be made to reach agreements on the kinds and degree of authority that can be exercised and whether it should lie in the hands of individual states, groups of coastal states, groups of coastal and distant-water states, or some forms of regional or international bodies.

IV Alternative Institutions for Management and Distribution

WHATEVER APPROACH is adopted—species, zonal, or combinations or variations of the two—it is clear that there will be a marked increase in the need for multinational arrangements. This will call for new institutions, or dramatic changes of old ones, in order to fulfill the various functions required for the management and distribution of the seas' wealth in fisheries. There is an infinite variety of the ways in which states may wish to organize their institutions. They will have to make choices as to the kinds of governing boards or bodies, voting structure, administrative mechanisms, means for financing the institutions, conduct of research, and a host of other matters relating to the acquisition and disposition of authority.

It is beyond the scope of this paper to discuss the many and varied forms of institutions, but it is appropriate to discuss the functions of management and distribution with which the institutions will have to deal, and to examine some of the alternative ways in which the functions might be fulfilled. Whatever form of institution emerges should be able to deal adequately with each of the functions and be sufficiently flexible to respond to changing developments and pressures.

The most important functions for the institutions are: the provision of data and information; regulation of the fisheries; enforcement of the regulations; and the settlement of disputes. As stated in Chapter III, it is important to adopt techniques for management and distribution that help to minimize the costs associated with the fulfillment of these functions. This should be kept in mind during the following discussion of the objectives of each function, the present methods for fulfilling them, and the alternatives for the future.

PROVISION OF DATA AND INFORMATION

This function includes the conduct of research, the gathering and compiling of data and information, and the provision of such data to those who must make decisions regarding management and distribution of living marine resources.

Goals

The initial question to consider is what objectives should be sought in the performance of the function. The following goals are suggested.

Development of Sufficient Data. It may seem unnecessary to suggest that an institutional device or arrangement for management should be able to supply decision-makers with the information needed for wise decisions. The point here, however, is to emphasize that the system should not be required to provide all the information it is theoretically possible to generate or acquire regarding a particular subject before the decision-maker can proceed with his task. At the same time, the system should not be content with acquiring a sprinkling of information which is sufficient simply to establish a *prima facie* case for a particular proposition. The former situation would mean, given the enormous complexity of the subject matter involved, that the decision-making process might be paralyzed. In the latter instance there is a considerable potential for reaching irrational conclusions because of inadequate information. The goal is to establish a reasonable balance, one in which the information adduced permits conclusions without excessive doubt attached to them even though some vestiges of reasonable doubt may remain.

Timeliness. A major desideratum for a fishery management institution is the acquisition of information in time to act effectively. In part this goal is affected by the amount of data that is demanded by the participants, and in part by the rapidity with which current developments in fisheries are taking place. The apparent ability of large fleets of vessels to deplete a stock in a few seasons places heavy burdens on the research function. It necessitates quick responses in potentially damaging situations and a willingness on the part of the participants to accept the minimal amount of information sufficient to make reasonable decisions.

The provision of information in a timely manner can be significantly influenced by the level of investment devoted to developing an adequate information base for decisions. Unless those concerned are willing to expend the resources required, it is plain that the research function will be hampered and that it will be difficult to do the work required to get the information desired.

The capacity of the institutional framework in terms of response time is influenced also by the traditional methods for gathering data.

There are very few instances in which data about a fishery are gathered on the international level. In most instances each state with vessels engaging in a fishery attempts to gather data about fish stocks and the activities of its nationals. Even though each national system for acquiring, processing, and storing data may be excellent, these may not be entirely compatible with each other, and considerable time may be required to collate and conform the data. In such circumstances it is not easy to build a research arm which employs data generated on the national level. This situation exists, and it is a serious obstacle to establishing better institutions for research pertinent for management.

Reliability. An important consideration in examining alternative institutions is that the conditions of operations facilitate the production of reliable data. This requires adequate staffing of research agencies with persons possessing the necessary training and education for the scientific and technical tasks. Again, this calls for appropriate material support for research consonant with the magnitude and complexity of the tasks involved. This latter consideration is by no means trivial, since the present level of investment for research in fisheries management is generally inadequate in most nations. There is often a noticeable and unfortunate contrast between the sums devoted to fisheries development and those for management, including research.

These various concerns gain in importance and emphasis when it is considered that institutions must now frequently deal with the problems of distribution as well as those of management. The level and kinds of research into the economic, social, and political considerations pertinent to distribution decisions increase the requirements for human and material support.

Acceptability. It is enormously important that institutions for research gain a reputation for scientific accuracy and credibility and, hence, that the work product is acceptable to those affected by management decisions. It may be that scientific "truth" is relative but it is still exceedingly important that the procedures and the conditions for doing research contribute to both the reality and the appearance of objectivity and impartiality in the process. Failure to attend to this aspect of institutional creation may undermine the entire management system. Previous experience has indicated that this problem is not insurmountable.

Minimization of Costs. As noted, different kinds of management and distribution techniques can bear significantly different costs for research. But, whatever the techniques, research costs can be considerably reduced by avoiding redundant efforts of national governments and facilitating cooperation and coordination in the acquisition and provision of information.

Research generally requires high investments in capital and scientific manpower. It makes use of costly specialized vessels, laboratories, and equipment for the collection, processing, and storage of data. It also requires highly trained scientists and technicians. Where each nation insists on providing its own information, because of an unwillingness to accept that of an international or regional body, this not only limits the timeliness and effectiveness of the information, but also leads to great duplication of effort and redundant costs.

Present Methods

For the world as a whole there is no question that the primary responsibility for meeting the information needs of fishery management rests with the nation-state. However, this does not mean that some international bodies do not now play a role of some significance. Three international regulatory bodies have their own research staffs which undertake the scientific investigations necessary for the actions of these bodies. It is worth special note that two of these—the Halibut and Pacific Salmon Commissions—have had scientific staffs virtually from their inception, and are among the two oldest existing fishery regulatory bodies. The Inter-American Tropical Tuna Commission also has had a staff since its beginning in 1951. In addition, the International Council for the Exploration of the Sea (ICES) deserves special mention, since it provides scientific advice to a political body, the Northeast Atlantic Fisheries Commission, but manages to maintain itself as a largely independent scientific organization. However, with these very few exceptions, the other regulatory bodies are essentially limited to a coordinating role for the research efforts of the individual member states.

The largest and strongest of the international bodies concerning fisheries is, of course, the Food and Agriculture Organization of the United Nations. The FAO has itself no regulatory functions but it has especial importance for its research and information activities pertinent to both fishery development and regulation, national and international. On a worldwide basis, certainly, there is no doubt that FAO occupies a unique niche as a depository and purveyor of statistical information pertaining to national fisheries. It has been useful in the provision of expert assistance to states wishing help in national fishery regulation. Its expertise has also been extremely useful to international regulatory bodies, not only for their own informational and coordinating efforts, as in provision of assistance on standardization of statistical data between or among members with different systems, but also in facilitating cooperation between various international bodies with common problems, such as ICNAF, ICES, and NEAFC.

Generally speaking, existing institutions are active in management efforts rather than in distribution of wealth, hence they generate information mostly for the former purpose. Even where the international body plays a role in distribution, the data for the decisions come from national rather than international agencies.

As a general conclusion, the present methods for research are inadequate and unsatisfactory in meeting the new pressures on fisheries and fulfilling the new requirements. First, the rapid growth in highly organized, mobile fishing effort greatly increases the need for speed in the provision of information and for improvement in the information base upon which decisions can be made. Second, the responsibility for making decisions on management is growing rapidly as national jurisdictions are extended and as new international arrangements are being made. For example, the proposal for controls on effort in the Northwest Atlantic creates a need for economic information that is not now available. And third, the necessity for dealing directly with distribution problems also places new pressures on information systems.

Alternatives

The various institutional methods for coping with the research function include (a) reliance on the coastal state in a greatly enlarged area subject to coastal jurisdiction over resources; (b) reliance on international regional agencies such as those now existing but with greater decision-making authority; (c) the creation of an international research agency whose competence would extend to all the world's fisheries; or (d) some combination of the above. An assessment of these alternatives follows.

Coastal State as Primary Research Agency. In one sense, future reliance on the coastal state for the research function in fishery management would simply be a continuation of past practice. In another sense, however, enlarged areas for coastal jurisdiction introduce such different problems that mere extrapolation of previous arrangements should not pass without question. The greatly increased range, number, and kind of problems suggest that the coastal approach cannot adequately cope with this task in many situations.

Relatively few coastal states are really capable of the overall management role, including that of undertaking research. Few states have either the cadres of specialized personnel who could do the research

66

required for management or the necessary vessels and equipment. Even the United States, rich as it is, would have difficulty in establishing a resource management effort of the requisite size and complexity. Numerous other states simply would not be able to discharge this function without considerable outside technical assistance. Reliance on the coastal state alone would, in some circumstances, lead to failure to develop sufficient data for conservation, cause failures in the provision of timely data, result in diminished reliability, and, for these reasons, diminish the acceptability of the research results.

At the same time it is not entirely necessary that the coastal state which discharges other management functions should also undertake the research activity entirely on its own. Apart from the potential input from international regional organizations, it is possible that a group of states might band together to provide mutual assistance in accomplishing research objectives. Another device might be to establish a global pool of experts whose advice would be made available to coastal officials as needed. Either of these approaches might be of major assistance in strengthening the coastal state approach and ameliorating the disadvantages noted above.

International Regional Institutions. The advantages of using multilateral agencies organized on a regional basis vary depending on whether the agency is to undertake the research itself or is to act as coordinator of national efforts. As noted earlier, there are examples of both these approaches in current practice, and it is generally believed that the international research staff is the better alternative.

In either event there are marked benefits from the international approach, the most obvious being that a cohesive international agency virtually assures that research efforts will extend to the total range within which the stock is fished, and permit much more effective assignment of research effort than is possible from an individual coastal state approach. A potential disadvantage of the international approach is that, if past experience is repeated, an independent research staff will be inadequately funded in comparison with the sums allocated to national agencies. This, however, might well be balanced by the large savings that could be effected by avoiding redundant research efforts.

An international regional research operation employing an international staff may have greater general acceptability than the research efforts of any particular individual state. Such a staff need not be recruited only from nationals of the coastal and fishing states concerned; there are considerable benefits obtainable from recruiting research personnel on a worldwide basis.

Although it is not an inevitable corollary, it seems likely that an international research staff might be better able to meet the problems of timeliness and sufficiency of data, particularly the latter. There is no inherent reason, of course, why coastal officials must be deficient in either respect. However, where scientific information is drawn from more than one state, there is an understandable tendency to insist upon higher standards of proof for the propositions the data are used to support.

The international regional approach is attractive also because it facilitates the solution of the problem of incompatible data systems. If coordination is to be effective at all, the states concerned have at least to adopt similar procedures and methods for data handling. Where each coastal or fishing state proceeds independently to develop and to maintain a data system, there is no assurance of compatibility and considerable likelihood of incompatibility.

A Global Research Agency. One possible advantage of a centralized research agency for the world's fisheries is that it could bring to bear the required expertise for developing the information necessary for proper management. This assumes, of course, that states are agreeable to providing the material support actually necessary for this task; an assumption of questionable validity.

There are some major disadvantages attending this approach. The overall magnitude of the task coupled with the necessity for understanding its numerous and varied individual components strongly suggest that a single agency would face an extremely difficult chore. The world fishing industry is, in aggregation, an enormous and diverse conglomerate of enterprises. It borders on complete impracticality to propose that one large research organization could adequately perform all of the information functions required to make management decisions in such a context.

A major difficulty would involve provision of information on a timely basis. In the fisheries as now conducted around the globe, developments of management significance can and do occur rapidly. A research apparatus must be able to respond with promptness. Unless the world agency were carefully structured and organized into decentralized units, it would be very unlikely that it could service management operations. This problem is not insurmountable, of course, but it does present severe difficulties in administrative arrangements.

Global scale agencies specialized to particular species do not seem subject to the same caveats, since they would operate on a far smaller scale and with a much more sharply defined mission. For such species as the tuna, for example, a global management agency with research responsibilities is likely to have considerable advantages over individual national or even regional efforts, since the species is caught on a world-wide basis by vessels with global mobility.

While a global scale agency may have limited value for performing the wide range of research functions necessary for specific management decisions, it can provide certain general kinds of information helpful to regional agencies and individual states. Obviously, the collection and dissemination of data on world catch and global fishing effort requires a global agency. This is also true for the important projections of the conditions of supply and demand, for information on changes in the ocean environment, the development and evaluation of new fishing techniques, coordination between regional and national research efforts, and a whole host of other matters of general relevance or worldwide significance. The FAO Department of Fisheries can continue to perform many of these functions and might usefully enlarge such activities. It might also facilitate specific research tasks by expanding its advisory services to regional organizations. But, while there is much that FAO (or some similar international organization) can do in this regard, the magnitude and variety of research tasks throughout the world is beyond the capability and budget of any single organization.

A Combination of Institutions. Clearly there are major advantages to be gained from the simultaneous employment of various institutional means, including the coastal state, regional, and global agencies. Fisheries problems are too complex and varied from place to place and over time to admit of a single management or institution approach. The question is, What combination of institutions is optimum for undertaking the research function? Previous discussion has suggested factors pertinent to an appropriate balance in this regard.

REGULATORY ACTION

Goals

The varied goals of the institutional method for laying down general rules of conduct include adequate geographic scope of coverage, flexibility in technique or method of action, the capacity to respond in a timely fashion to the workings of the fishery, and wide acceptability among those affected by regulation.

Geographic Scope. It is elementary that regulatory measures must be applicable throughout the range of the fishery or fisheries sought to be managed, if there is to be any prospect for effective action on the conservation side. Accordingly, this goal is virtually the *sine qua non* of a regulatory regime that has any substantial prospect for success.

Flexibility. The rate of change in problems confronted in fisheries management makes it essential that fishery institutions be authorized to employ whatever management or distribution technique appears able to do the job. One of the primary difficulties in previous approaches has been that the regulatory agencies have not been authorized to take the full range of actions that turned out to be essential for maintaining viable fisheries and satisfactory political relations among the fishing states concerned.

Timely Response. Management institutions must be designed that are capable of making decisions without long delays and repetitious consideration of pressing issues. The distribution of authority between a multinational agency (if such is to be employed) and member states must be structured so as to allow the agency a realistic opportunity to take the necessary decisions. This is an extremely important factor for effective regulation but it is difficult to arrange since it implies that the agency must be able to operate without the unanimous concurrence of its members on all issues. The significance of this objective is only heightened by the rapidity with which changes of regulatory significance can and do take place in the fisheries.

Acceptability. It hardly needs to be emphasized that an overriding objective for any management unit is that its prescriptions command sufficient acceptance among those affected to be effective without continuous supervision or surveillance. It is one thing to expect that there will be occasional evasions of general regulations but it is quite another to have massive disregard for authoritative rules. Accordingly, the institution for regulation must be one which ensures uniformity and even-handedness in dealing with problems, and which commands sufficient confidence that its actions will generally be respected by those subject to them. This strongly suggests that the unit must possess the necessary skill and competence to undertake knowledgeable management actions. This is not a trivial detail since, in many parts of the world today, one might well doubt that there are adequate cadres of personnel who possess the necessary training and education to equip them for the complex tasks of fishery management.

Present Institutions for Regulation

Although the numerous international fishery bodies are often said to exercise regulatory powers, the fact is that they seldom have the

70

basic authority to actually prescribe regulations that directly bind member states. With some minor exceptions, the fishery commissions are limited to making recommendations. The FAO Report on Regulatory Fishery Bodies, presented to the UN Sea-Bed Committee in March 1972, accurately and succinctly notes the restricted authority these groups are given:

> Regulatory fishery bodies do not possess supra-national powers and the conservation measures they formulate and adopt are not directly binding on individual fishermen without legislative action being taken to this effect by member countries. In fact, these measures are seldom even binding on member countries. However, in a few cases, conservation measures are automatically binding on member countries. This occurs only within the framework of regulatory bodies which have a limited membership and in which the decision to adopt a given measure requires the unanimous vote of the member countries. Thus, the North Pacific Fur Seal Commission may determine the total number of seals which may be taken at sea for research purposes as well as the number to be taken by each member country; the Mixed Commission for Black Sea Fisheries may adopt measures concerning the species and dimensions of fish that may be caught in the Black Sea; the International Pacific Salmon Fisheries Commission may issue orders for the adjustment of closing or opening of fishing periods and areas in any fishing season as well as emergency orders required to carry out its functions; the Japanese-Soviet Fisheries Commission for the Northwest Pacific may fix the total annual catch of a stock of fish and determine the annual catch of such stock by each member country. (pp. 12-13.)

In recent years, however, there have been attempts to strengthen the recommending function of some commissions. A number of the international bodies now have provisions in their basic charters to the effect that if member states do not object to a recommendation from the commission it becomes binding on them. This represents a reaction to the previous unhappy experience under provisions that made the effectiveness of recommendations dependent on the affirmative response of all members. In such instances, even careless delay or neglect of the body's actions frustrated its attempt to respond to a problem needing international attention. The FAO Report (p. 13) describes this approach as follows:

> . . . member countries may lodge an objection to the recommended conservation measure within a given period (from three to six months as the case may be) of being notified of it. If no objection is made during that period, the recommendation becomes legally binding upon all member countries. If, on the contrary, one member

country objects to the conservation measure, the entry into force of the recommendation is postponed for an additional period during which the other member countries may reconsider their attitude in the light of this new situation. At the end of the additional period, the recommendation becomes binding upon all member countries that have not objected to it. If, however, objections are too numerous, the recommendation does not become effective, except that certain member countries may agree among themselves to give effect to it. It may be added that the Convention establishing the International Commission for the Conservation of Atlantic Tunas contains provisions whereby, in cases in which few member countries lodge an objection, their objections will be deemed to have been withdrawn unless they are reaffirmed within an additional period.

Even this method becomes operative only after the plenary authority within the regulatory body has adopted a recommendation, and sometimes this must be done by unanimous vote. In many instances, member states are represented on the regulatory body by the same officials who must later make the effective decision whether or not to accept a recommendation. Accordingly, an affirmative action to make a recommendation means, in practical effect, that it will become binding because no member state is likely to object to it.

The reason fishing states are reluctant to confer genuine supranational authority on a fishery agency is, of course, not hard to imagine. There are very few instances in any field of governmental activity in which states confer both formal and effective power upon an international institution to permit it to take actions independently of the members' review for ultimate acceptance. Government officials usually do not advance in office by agreeing to relinquish their powers to others. Perhaps as basic is the fear and mistrust of others that still is generally symptomatic of the entire international system. In the specific matter of fisheries, there is the additional consideration that financial support for management efforts is by no means plentiful. Affected national fishery management officials are not anxious to create new and conflicting demands for scarce funds. This consideration also partially accounts for the reluctance of nations to provide new regulatory bodies with a research staff, despite the very considerable success existing agencies have had with such a staff.

As noted elsewhere the limited authority of fishery regulatory bodies has been confined mostly to recommendations regarding conservation measures, with but little attention being authorized for consideration of distribution problems. There are a number of fishery conventions and commissions, however, which do play a role in this respect, including the Fur Seal Agreement, the INPFC, the Salmon Commission, the Northwest Pacific Commission, the LATTC, ICNAF, and the

Whaling Commission. Of these groups, however, only ICNAF can be said to involve a multistate, multispecies fishery, while the others are predominantly concerned with but a single species or only a few states.

Alternatives

It is not at all difficult to conceive of the general nature of the alternative institutions that may be devised to undertake regulatory actions. But it is extremely difficult, and is not here assayed, to conceive of the infinitely varying details that might attend regulatory systems around the globe. This task is better attempted in specific contexts, and is, therefore, properly a component of the various regional studies that are being conducted for this series.

With the possible exception of certain highly migratory oceanic species, the institutional mechanisms for regulation of fisheries all involve coastal states to a greater or lesser degree. The institutions range from exclusive coastal state jurisdiction, unilaterally acquired, to a mixture of jurisdiction with other coastal, regional, and/or distant-water states. Variations are such that, even a management system depending mainly on coastal authority may have many permutations, ranging from complete authority in that state to varying forms and degrees of limitation on such authority. Establishment of coastal state jurisdiction through unilateral action generally involves an assumption of complete authority. The same arrangement might also arise by international agreement, although it is possible that negotiations at the Law of the Sea Conference could result in some limitations on coastal state authority.

A major advantage of relying on a coastal state approach is that it may offer the potential of a management unit with both the necessary geographic scope of authority and the political authority required to make effective decisions. The first of these objectives is possible, though in many fewer instances than is frequently assumed, where extension outward of coastal jurisdiction for fisheries includes all the range within which coastal stocks move during the harvestable stage of their life cycle. The second objective follows because the coastal state need not obtain the consent of any other political authority before taking management action. This latter circumstance also has the advantage of stimulating realistic negotiations with states wishing to fish in the coastal jurisdiction, since the *only* way to gain access to such an area is by agreement with the coastal state.

The coastal state system also permits other objectives to be accomplished. Thus the coastal state has a high degree of flexibility because it may utilize any management that it wishes, there being no need to adopt these subject to the concurrence of other states. Similarly the possession of sole authority to make decisions means, in practical terms, that coastal state officials can act with promptness and dispatch as a situation may demand. There is no need to negotiate with other units to gain their concurrence—negotiations which can and do require time and effort inconsistent with moving quickly as problems arise.

There are, however, some major difficulties associated with exclusive coastal management. First, the extension outward of coastal jurisdiction does not at all assure that the coastal state's authority actually encompasses the entire range of migration of the stocks. Even if coastal jurisdiction is *defined* in terms of species, as some states suggest, the fact that fish also move laterally along a coast, as well as in and out from deep to shallow water, means that the range may extend into the jurisdiction of the laterally adjacent state. When this occurs, neither can be assured of being able to exercise exclusive jurisdiction at all relevant and necessary times. In such circumstances there may be no alternative to a bilateral or multilateral institution for management.

Another shortcoming of the coastal state approach is, in the very likely possibility noted above, that an individual coastal state (or even group of such states) does not possess the indigenous skills and training required to make knowledgeable management decisions. The significance of this is twofold: it suggests, first, that management decisions may not be adequately informed and therefore fail to achieve the ends sought, and second, that, as a result, there may be substantial difficulty in achieving acceptance of the decisions. It should be added that this difficulty is not insurmountable. It might be possible to establish an international or regional mechanism for assisting coastal states to perform management responsibilities until such time as they can do so on their own. Such a mechanism would, of course, add to the cost of management for the period it is needed.

The assumption of enlarged coastal state authority may, in some situations, have a detrimental effect on the production of values from fisheries. It is not always in the best interests of a coastal state to invest heavily in fishery industries. Such an investment may divert scarce capital and manpower away from activities that could make much greater contributions to the nation's economy, and it may have little vaue in the production of food, income, or useful employment opportunities. In such situations, and where there are high value stocks in the coastal waters, the prohibition of all foreign fishing would diminish the values that might be produced from the stocks. This is unlikely to occur to any great extent, because most states would find it to their advantage to collect revenues from foreigners for fisheries that they themselves cannot fully utilize. And they will find that attempts to impose excessive taxes or license fees will serve to diminish their returns. But in some cases, perhaps only temporarily, undesirable prohibitions might occur. Where this takes place, it could, in addition to reducing values, conceivably lead to serious conflict. States with large, specialized, distant-water vessels might find that the values of violating coastal state authority are worth the risks they bear. The greater the value of the stocks and the weaker the capabilities of coastal state enforcement, the more likely this is to occur, and no one will benefit.

Another contingency that needs to be weighed in assessing this alternative is whether coastal state jurisdiction can, in fact, avoid the international negotiation that seems to be the hallmark of multilateral approaches. Paradoxically, the extension of coastal states' authority may lead to greater rather than less reliance on international bodies for assistance on regulatory measures. If greatly extended fishery jurisdiction were generally accepted by treaty, or through customary methods as in the past, there would be, of course, no doubt that coastal states could lawfully make their own decisions. However, for somewhat the same reason as noted above regarding research and information, many coastal states may wish to look to other sources for either recommendations on such actions or for independent decisions. States whose scientific institutions and personnel are insufficient for research into living resources may also rationally prefer that the expertise afforded by a multistate agency be utilized for translating scientific and other information about a fishery into courses of action for regulation.

Justifications for use of a regional or international body may occur where extensions of jurisdiction disproportionately favor one coastal state and leave neighbors without access to comparable quantities of marine resources. These one-sided situations could lead to demands, as in the Caribbean, that the states of the region share in the authority and in the proceeds or benefits of a fishery which would otherwise largely be controlled by a small number of the states, if all were to recognize each other's expanded resource zones.

These various situations indicate that the net result will be a wide variety of institutions for fishery regulations. At one extreme, full authority for regulations might be assumed by the coastal state. This might occur where individual stocks are fully enclosed within its jurisdiction, and the state is competent and willing to take on the tasks of regulation. At the other extreme, there may be little or no authority vested in coastal states, and regulations might be adopted and imposed by an international body. This form of institution might occur where stocks lie wholly or largely outside of national jurisdictions and where depletion has been severe because of past ineffectiveness in regulations. If values have been so severely diminished that no individual user state has much of an interest in the stock, and if there are no coastal states that can claim an interest, it may then be possible to confer a high degree of authority on the international institution.

Between these two extremes will fall the majority of regulatory bodies, binational or multinational in organization, with varying degrees of authority. In order to achieve flexibility and timely response to changes in fisheries, and in order to adopt the most effective regulatory techniques, the institutions should have as high a degree of authority as possible. That is, they should be free to adopt, change, and impose whatever regulations are available and desirable to permit the greatest production of values from the fishery. This will be facilitated if the functions of regulation can be separated from those of the distribution of values. However, while there seems to be some indication that new multinational institutions have been able to acquire a greater degree of authority than usual, this is only a slight improvement and mostly restricted to strengthening the recommending functions of the institutions. It seems rather likely that these bodies will continue to be limited to recommendation of measures rather than direct prescription of them. This seems to follow from the very fact that states generally are not now inclined to rely on international fishery bodies for effective prescription, or they would not need to expand their own jurisdiction. Once they have accomplished the latter course of action, if it is accomplished, it would be surprising if coastal states relinquished much of their authority to another agency.

ENFORCEMENT OF REGULATIONS

Goals

Specific goals for enforcement include efficiency and fairness in application of regulations and building a sense of trust and confidence in enforcement measures and procedures. These goals are not entirely separable.

Efficiency and fairness involve a system that operates without excessive cost throughout the management area, and does not discriminate unreasonably among fishermen, by national character or for any other reason. Effective law enforcement requires a surveillance force that has sufficient size and scope of operation to apprehend violators but does not require disproportionate expenditure of time and resources. There should be no privileged sanctuaries for a certain class of fishermen or boats, nor should there be a pattern of enforcement that weighs disproportionately against a particular region or class of boat. Obviously, these aims have some specific connotations for institutional mechanisms.

76

An essential element of effective overall management is that those regulated believe that offenders against regulations will be brought to justice promptly and that prescribed sanctions will be imposed without delay. Unless those subject to a regulatory management have some minimum confidence in the administration of the system, they may disregard prevailing regulations, perhaps in large numbers, and reduce the whole management scheme to futility. Even if mass violations of regulations do not occur, the incidence of individual offenses may rise and make enforcement virtually impossible. It is, accordingly, essential that the enforcing agency not only behave without discrimination but that the agency *appear* to operate this way.

Present Methods of Enforcement

For the sake of clarity in understanding, it would be ideal to discuss the total process involved in the application of fisheries regulations, including the creation of the system of application. For present purposes we need refer only to the more important components of the process, including the decision to initiate or to change an enforcement system; the various phases in the operation of the system including surveillance, arrest, and trial; and the means for appraising effectiveness of the system. In general, the role of international bodies in any of these stages of the decision is minimal, and the nation-state is still easily the most influential participant.

Decisions to Establish or Change an Enforcement System. Although numerous international regulatory bodies make recommendations for regulatory action concerning fisheries and the constitutions of several make provision for enforcement, only three commissions are presently authorized by their constitutive charters to make recommendations of enforcement. Although some other commissions may discuss enforcement problems or even take some action to make a system operate more effectively, it is manifest that most nation-states have been unwilling to confer any initiative on their representatives to act collectively, either to recommend establishment of an enforcement system or to propose modification of an existing system. There is, perhaps, no better illustration than this of the reluctance of nations to authorize even very modest collective action by institutions they have created, and wholly control in most instances. Thus, if the member states who agree on the various components and powers of an international regulatory body should later discover that the enforcement measures they provide are not working properly or must be changed to cope with changed conditions, they cannot act to recommend the changes. This means that action has to be initiated by governments through amendment of the agreement establishing the regulatory agency. It is very frequently, and sometimes wisely, made more difficult to amend agreements than it is for the agency itself to take recommendatory action pursuant to them.

Phases of Enforcement System. In applying the general rules adopted for enforcement of regulations, there are a number of steps involved. There is, first, the physical task of surveillance, which requires the use of vessels or aircraft actually present in the fishing area. In every such instance known, the process of surveillance is carried out by national officials in vessels either especially designated for the task or a part of a national enforcement service. As noted below in reference to arrest of vessels, this stage of enforcement is sometimes carried out by vessels of one flag in reference to those of another. (The Whaling Enforcement Agreement envisages use of the whaling vessels themselves for this task, with observers nominated by national governments engaged in whaling and selected by the Whaling Commission.) In no instance is there an international surveillance system in operation.

Unlike the arrest and trial phases of enforcement, there are considerable differences in the kind of surveillance required, associated with the different kinds of regulatory techniques. Closed areas for all kinds of fishing are probably the easiest technique to enforce, since visual observation from air or surface of the presence or absence of vessels is all that is required. If the area is large, however, or if other kinds of fishing can be permitted in the closed area, then the costs of surveillance increase. In the former instance, it means that a larger area must be covered; in the latter, there must be sufficient observation to determine the kind of fishing that is taking place or the kind of fish being caught. Regulations that require inspection of gear (controls on mesh size, for example) or inspection of catch (national quotas) create more difficulties for surveillance. These may actually require on-board inspection if they are to be effective. In some cases, where incidental catch controls are imposed (as, for example, where post-season catch of a regulated species cannot be more than a certain percent of the catch of unregulated species), surveillance may only be possible at the time when the catch is landed. These different kinds of regulations carry implications not only for the costs of enforcement but also for the kind of enforcement institution.

A second task, the arrest phase, is that of determining that a vessel has violated the regulations and apprehending the offender. This may be done either on the dock at the time of landing by designated inspectors, or by officials acting at sea. In either event the authority in every instance is conferred upon a national official and vessel; as noted,

78

there are no international agencies or personnel authorized to take any action except under the whaling accord. However, agreements differ fundamentally in their provisions in this respect. Some agreements authorize this step to be taken by the vessel of any member state with respect to the fishing vessel of any other member. Other agreements, in contrast, authorize this step to be taken by an enforcement vessel only against its own flag vessels. Clearly, these procedures differ in basic substance and are very likely to be perceived quite differently by those potentially affected.

A third step in application is that of actually trying the suspected offender and applying sanctions. In every current enforcement system, this stage is left to the state to which the offending vessel belongs, whether the actual arresting vessel is of its own flag or that of another member state. In the latter instance there are sometimes agreed-upon provisions concerning assistance to the trying state by making witnesses and evidence available. There may also be obligations on the part of the trying state to report to the fishery commission concerned on the action taken and penalty imposed. In some cases, the regulatory bodies review these national enforcement actions.

Appraisal. The final phase of application is that of appraising the system in use for achieving the goals set for it. The last-mentioned procedure—reporting on the actual sanctioning practice—provides information to regulatory bodies and member states for assessing the effectiveness of this aspect of the process. But since the international commissions have, generally speaking, very limited authority to do anything about the established system, and no authority to change it, improvements of any phase of the system can only be made by states acting individually or by amendments to the constitution.

Alternatives

The following discussion pertains only to the three phases of the enforcement process described above: surveillance, arrest, and trial and sanctioning.

Coastal State Enforcement. One alternative is that, by explicit agreement or otherwise, coastal states gain exclusive authority to manage fisheries in an extensive fisheries zone. If this authority is unconditioned by international criteria or standards pertaining either to management or distribution, it can be assumed that the coastal state will also bear the total burden for all phases of the enforcement process. The costs of the process will be borne entirely by the coastal state if it prohibits all foreign fishing. In some instances, the costs of attempting enforcement may be greater than the benefits that can be obtained. Alternatively, the coastal state may permit foreign fishing subject to the payment of a fee, which may help to defray enforcement costs. Benefits to the coastal state in this case will depend upon the amount of the fee and the kind of regulations that are being enforced. As noted above, different regulatory techniques may involve quite different kinds of enforcement measures and costs.

It is possible, of course, to imagine other arrangements involving coastal state enforcement. For example, if the coastal state were authorized to apply *international* regulations to all vessels in a fishery (i.e., conduct surveillance, make arrests, conduct trials, and impose punishments), it might then be possible, and certainly advantageous, to allocate the overall cost among all states participating. Another possible arrangement is to assign various phases of enforcement differently. Surveillance might be assigned to one group, arrest to another, and trials to a third. It is common, as noted above, that however the first two of these tasks are performed, the third is always retained by the flag state.

A major difficulty with any system that relies on the coastal state alone (or on any single fishing state) is the suspicion, bound to arise, that enforcing authorities are partial to their own flag vessels. Still, this arrangement may be more attractive than flag state enforcement, due to the expense of the latter approach. A progressive dispute settlement procedure would lend even more attractiveness to such a system.

Flag State Enforcement. The role of flag state vessels and officials in enforcement depends upon the degree to which the fishery is shared, the kind of regulations to be enforced, and the willingness or unwillingness of states to permit their fishermen to be subject either to coastal state or international enforcement vessels. With regard to the phase of surveillance in shared fisheries, it is difficult to avoid the use of flag states where the regulatory device is such that inspection of landings is required. On the other hand, where aerial or surface observation is sufficient, insistence upon the use of flag state aircraft and vessels will lead to wasteful redundancies. In this case, it would seem preferable to confer the requisite authority upon a single state or an international agency, with costs in both situations shared by participating states.

For arrest, if it is required that each state employ its own vessels, there would also be redundant costs, in that each state would have to have arresting vessels in the area to serve this function. Here, the primary question is the degree to which the member states are willing to concede authority for arrest to others. This depends, in part, upon the degree to which they accept the fairness of the arresting agent, whether that agent is a single state or acting for all member states together.

As noted before, it would be very difficult in the case of shared fisheries for states to agree that trial and sanctions of their fishermen should be outside of their control, even though this may merely substitute one system of biased judgment for another. In all of these cases, some form of compulsory dispute settlement is likely to make it easier to avoid the excessive costs associated with flag state enforcement.

International Enforcement. As indicated above, a wholly international institution for all phases of enforcement is unlikely to be acceptable for many years to come, if ever. No international body now exists that has a right to bring nationals or nation-states to trial, and to impose sanctions upon them. International institutions for surveillance and arrest are more readily conceivable and would, where in effect, produce savings for the member states. In addition, once they are accepted, they may well be able to achieve a higher degree and sense of fairness than exists where these functions lie solely in the hands of flag states.

DISPUTE SETTLEMENT

Goals

The establishment of an impartial, objective procedure for making decisions in disputes over management or distribution is of enormous importance to resolving genuine differences about the law of the sea. It may be no exaggeration to say that provision of such a procedure is the key question to be resolved at the UN Conference. Without third-party decision over disputes, several states will find substantial difficulties in concurring with extensions of coastal state authority.

A key difficulty is how to arrange a dispute settlement procedure that is virtually automatic, i.e., beyond the control of a state party, and timely. Both of these elements are essential to successful creation of an acceptable institution. The more difficult of the two is perhaps the latter, since attempts to avoid control so often involve time-consuming procedures.

Currently, when disputes arise over fisheries, states employ the normal methods developed to manage their differences; namely, negotiations and ad hoc agreements. The recent growth in exclusive fishing zones is, not surprisingly, accompanied by an increase in bilateral agreements, adjusting relations with nations affected by the new zones. This method is obviously preferred to judicial settlements. The recent and still current dispute between the United Kingdom and Iceland over resort to the International Court of Justice, pursuant to their 1961 bilateral fisheries agreement, illustrates attitudes in this type of dispute. Iceland's rejection of the Court's jurisdiction is indicative of the general attitude that judicial settlement is unlikely to be acceptable where the edges of conflict are still very sharp.

In the high seas beyond coastal state authority, it is also unusual to find that a group of states, or even two states, have agreed upon a special procedure for settling disputes over fisheries. The only notable multilateral agreement containing such provision is that concluded at the 1958 Conference, in the Convention on Fishing and Conservation of the Living Resources of the High Seas. As is well known, this Convention has not attracted wide adherence. One of the reasons why only a tiny segment of major fishery nations has ratified it, is the nature of the provisions on compulsory settlement of disputes.

The Convention contains a variety of provisions according to which a state may be subject to conservation measures adopted by other states, including a coastal state. In the event of disagreement over such measures and their nonacceptance by an affected state, any state may invoke a dispute settlement procedure which utilizes a special commission that can operate without the consent of the states directly concerned. Although this arrangement makes a considerable advance toward impartial third-party adjudication, it has not been used. So far as is known, the 1958 Convention has never formed the basis for any conservation action, and there has been no occasion for employing the dispute settlement provisions.

The numerous international conventions which established fishery commissions almost never contained provision for settlement of differences arising in implementation of the agreement or in the operation of the commission. The 1957 Interim Fur Seal Convention provides in Article XII for consultation of the parties where any party considers that an obligation of another party is not being carried out. However, this arrangement is rather cursory, for if "such consultations shall not lead to agreement as to the need for and nature of remedial measures, any Party may give written notice to the other Parties of intention to terminate this Convention . . . ." And, without further action, the Convention terminates as to all parties in nine months from the date of notice.

The 1967 Convention on Conduct of Fishing Operation in the North Atlantic is quite unusual, since it provides for arbitration of disputes concerning the interpretation or application of the Convention. The degree of this departure from usual procedure is suggested by the fact that the article on arbitration is singled out as the only complete article to which any state may make reservation.

The major change in fishery arrangements that seems virtually certain to transpire, whatever is or is not achieved at a future law of the sea conference, is the necessity for dealing directly with problems of distribution of fishery resources or benefits from fishing. These decisions could, by international agreement, be assigned in many cases to the prerogative of coastal states for resources within a large nearby region. Perhaps too, though this possibility is not at all clearly foreshadowed, there might be provisions made for assigning distribution decisions to groups of states, including one or more coastal states that might or might not actually engage in fishing. Whatever the precise alternative negotiated at the conference, states might see common advantage, in some situations, to establishing different institutions for making management and distribution decisions.

The point of this observation is simply that there appears to be a very widespread agreement that states must expressly and directly deal with the question of who gets what, rather than continue either to ignore the matter or to cloak decisions about it in the guise of conservation. If nations generally act to make the problems of distribution a primary one in their relations, the prospect is that there will be disputes over the creation or application of particular formulae for governing the matter. This would seem especially likely if there is to be a widespread agreement on extensions of coastal state jurisdiction, since such extensions affect not only distant-water states but also, and perhaps more importantly, the interests of neighboring states sharing a common resource. In order to get general agreement on this approach, it may be important to assure that decisions purporting to comply with certain standards shall, on challenge, be reviewed by an impartial tribunal.

Accordingly it would appear that in the evolution of an improved international institute for fisheries, one extremely important ingredient is a dispute settlement procedure that works. Except for negotiations to resolve disputes, there has been no agreed and effective system prior to this time. One point worth emphasis is that there need not be only one universal institution or procedure devised—the problems of fisheries, including that of relations between nations, vary around the globe, and in taking this into account there may well be more than one approach to the problems.

V General Principles for Fisheries Arrangements

INTRODUCTION

SOCIETY, like nature, abhors a vacuum. It cannot prevent jurisdiction from flowing into the vacuum where valuable resources are subject to no one's control. The greater the value of the resources, the greater the pressures to fill the vacuum, and the more rapidly the flow of jurisdiction of one form or another. Once the resources reach a certain value, the process becomes inexorable.

This stage has now been reached for many of the resources of the sea. Until recent years, there was little value in having control over high seas fisheries. It was sufficient to operate under the principle of the freedom of the seas and under national jurisdictions which were limited in both content and scope. But the value of fishery resources has now become greatly increased, due to the pressures that a large growth in demand has placed upon severely limited supplies. It has now become inevitable that jurisdiction will flow in to govern the use of most marine fisheries.

Society is, thus, facing a period of transition, during which the principle of the freedom of the seas for fishing will be dismantled and replaced by the imposition of national authority—an authority that might be exercised either unilaterally by a single state or jointly by a group of states. The task is not to impede this transition, which would be both fruitless and dangerous, but to facilitate it in a manner that will minimize conflicts between states and minimize damages to man's interests in the resources.

There are two dimensions to this task: one is appropriate for consideration at the UN Conference on the Law of the Sea; the other is more properly the job of regional groupings of states. This paper has, thus far, attempted to cover both of these dimensions simultaneously, and, in doing so, may have made the problems of fisheries arrangements appear to be overwhelmingly complex. But in this final chapter, the dimensions are separated. The problems of determining the actual arrangements in particular regions and situations are not discussed further in this paper, but left to the more detailed presentations in the other studies in this series. Instead, the few remaining pages focus attention on the problems likely to be addressed at the UN Conference—those relating to the adoption of general principles that might be helpful in guiding the transition from the stage of no one's jurisdiction to the stage of someone's jursidiction.

GENERAL PRINCIPLES

The essential task at the UN Conference will be that of determining the scope and content of authority over marine fisheries. This is only in small part the question of the distance from shore to which a coastal state can exercise jurisdiction, because stocks of fish pay no regard to man-made boundaries and because some kind and degree of authority must be exercised over a stock wherever it is found. Instead, the question is one of achieving a balance between the authority of individual states and states as a group or as a whole. It is one of determining the locus of authority and the constraints under which it might be exercised; the kind and degree of universal principles that might be adopted.

Elements of this question are likely to be expressed in a variety of ways in reference to different kinds of situations. Some of the forms of universal principles that may be proposed at the UN Conference are identified and discussed below.

1. Should there be a separate regime for anadromous species of fish? If so, what should be the nature of the regime and how should the category be defined?

As noted in the discussion in Chapter I, there is some justification for according special treatment to the anadromous species, such as salmon, that spawn in fresh waters and swim in the high seas. The rationale for this lies partly in the fact that the yield from such species can be improved by investments in cultivation practices, and that such investments will not be made unless the host states can be assured of satisfactory returns. The rationale can also be found partly in the fact that the species can best be managed by harvesting it close to the mouths of the spawning streams.

These elements suggest the value of a regime that permits the host states to control access to the stocks wherever they are found, thereby prohibiting foreign fishing except with the host state's permission. They suggest also that the host state should receive a sufficient share of the catch, or sufficient benefits of other kinds, to warrant its investments in cultivation. But whether the host state should receive the *total* benefits is a question of wealth distribution rather than management. Although different techniques for distribution can be suggested (see Chapter III), there are no criteria other than that of acceptability, for providing an answer to this question.

If some general principles, similar to those suggested above, are adopted at the UN Conference, it will be necessary to arrive at an acceptable definition of anadromous species. This, as noted above (Chapter I) might be quite difficult, and it is questionable whether there will be sufficient time or expertise at the UN Conference to produce a definition that will be satisfactorily precise and generally acceptable. It might be preferable to give this task to a separate international body.

2. Should there be a separate regime for highly migratory oceanic species of fish? If so, what should be the nature of the regime and how should the category be defined?

It was suggested above (Chapter I) that highly migratory oceanic species, such as tuna, might also be accorded special treatment at the UN Conference. The rationale for this lies partly in the fact that such species are often found at great distances from shore, beyond present claims of jurisdiction. They are, thus, beyond the management competence of the coastal states. In addition, the mobility of many tuna vessels is so great that management measures in one area have significant ramifications for the use of tuna stocks in other, far distant, areas.

The suggestions for appropriate regimes to meet these special characteristics are the subject of a separate study in this series. (See Saul Saila and Virgil Norton, *Alternative Arrangements for the Management of Tuna,* Washington: Resources for the Future Program of Studies of International Fisheries, forthcoming.)

With regard to definitions, the same kinds of difficulties exist for this category as for anadromous species and the same suggestion might apply.

3. With regard to coastal species, several questions can be raised about the situation where a stock is found entirely within the waters of a single coastal state. In this situation, should a coastal state have a right to extinguish the stock? Or a right to deplete the stock? Under what, if any, conditions should a coastal state be required to permit foreign fishing for the stock? Or conversely, under what, if any, conditions can it prohibit foreign fishing? With regard to these questions, should the rights apply to the entire area of jurisdiction from the state's baselines on out, or only to the area beyond a distance of three, twelve, or some other number of miles?

The questions of extinction and depletion of a species might be treated separately from the other questions raised above. As noted previously (see Chapter II), the extinction of a species represents a loss to mankind as a whole. It destroys genetic material that can never be reproduced. Although extinction is unlikely in most fishery situations, because it generally becomes uneconomical to catch the few remaining individuals, the possibility exists. It may, therefore, be desirable to adopt a general principle prohibiting any individual state (or group of states) from fishing a species to extinction, even though the species occurs entirely within the territorial sea of a coastal state.

It should be noted, however, that such a principle may be little more than an admonition to behave. It is difficult to imagine a regional or international authority with the right to inspect the status of a stock found entirely within the waters of a single coastal state, and it is almost impossible to imagine one with the right to enforce conservation measures. Nevertheless, the principle may be desirable even though it cannot be made fully effective.

The question of the right of a coastal state to deplete a stock might also be raised at the UN Conference. As noted above (Chapter II), depletion of a stock may have a particular value to a state under certain conditions. On the other hand, some states may claim that depletion represents a loss in the supply of food available to the world community and that, therefore, no state has a right to deplete resources no matter where they occur. There seems to be little merit in such a claim. But even if there were, such a principle would be even more difficult to enforce than one prohibiting extinction, because of the difficulty in determining the extent of depletion and the costs involved in preventing it. Conversely, there seems to be little disagreement with the principle that a coastal state has a right to prevent depletion by foreigners.

It is more difficult to deal with the other questions raised about a state's authority over a stock that swims entirely within its waters of jurisdiction. To a large extent, these are questions of wealth distribution rather than management and not, therefore, amenable to discussion on the basis of rational, objective criteria.

If the stock is being fully utilized by the coastal state, it is likely that there will be general agreement that foreigners should not be permitted to have access to it. This principle might even apply under the species approach, which suggests that a coastal state should have rights to as much of an adjacent resource as it can harvest, even though the resource falls outside its limits of jurisdiction.

Where a stock is not fully utilized, or not utilized at all by a coastal state, a general principle on access by foreigners might be difficult to reach. It is hard to comment on this because of the wide variations in proposals that might be submitted. There are, for example, several possible interpretations of access. It might be suggested that access be free and open; that it be subject to a minimum fee to help defray expenses of administration; or that it be subject to as large a fee or tax as the coastal state desires. It might be suggested that access be available to all foreigners or only to certain selected foreigners, such as those from neighboring states (coastal or land-locked).

The variations in regional situations are so great that it is questionable that any general principle along the above lines would be universally acceptable. But more importantly, in terms of management and the production of benefits from marine fisheries, such a principle would hardly seem necessary. If a coastal state has fully exclusive jurisdiction over an unutilized stock, it is not likely to prohibit use of that stock when other countries are willing to pay for access. And, as noted throughout this discussion, the extraction of revenues does not necessarily either reduce the supply of food or lead to an increase in its price.

In spite of this, principles providing for free and open access, or for conditional access, by foreigners are likely to be proposed. These are essentially matters of wealth distribution rather than management and should be considered in those terms; for example, whether this form of wealth distribution is necessary in order to achieve acceptance of extensive limits of jurisdiction.

4. Where a coastal stock swims in the waters of two or more states, what, if any, general principles might be adopted?

Certain general principles that might be proposed and that relate to this situation are considered elsewhere. For example, a universal principle prohibiting states from extinguishing a species would have the same characteristics as those described in Point 3, above. Principles involving compulsory dispute settlement, international research services, and universal taxes, are discussed below.

With regard to depletion, however, the comments made in Point 3 may not entirely fit the situation where the stocks are shared. Here (as noted in Chapter II), depletion by one of the sharing states may be advantageous to it but damaging to the interests of the others.

Or, because of different goals sought from the use of the stock, effective agreement on management may be difficult to achieve. Under these conditions, the states may engage in a mutually destructive race to harvest the fish. They may want to consider, therefore, the adoption of some general principle to prevent the mutual loss of benefits. This might best be handled by a general principle advocating the creation of machinery for the settlement of disputes.

5. Should there be an international body that can provide coastal states and regional groups with research services? recommendations on regulations? imposition of regulations? enforcement services?

These questions are related to the degree of authority to be accorded to coastal states in that the authority depends, in part, upon the ability of the states to manage the resources within their waters. As noted in Chapter IV, the extension of jurisdiction forces coastal states to deal with problems of management. In order to ensure continued benefits from the resources, the states must have adequate knowledge about the stocks, adopt satisfactory regulations, and be able to enforce the controls. In many situations, individual states may find it difficult to fulfill these functions. It may, therefore, be desirable for the delegates at the UN Conference to consider the role of an international body in meeting some or all of these requirements. This is discussed more fully in a separate study in this series. (See Edward Miles, *Alternatives for International Fishery Institutions,* Washington: Resources for the Future, Program of International Studies of Fishery Arrangements, forthcoming.)

6. Should international machinery be created to facilitate (or compel) the settlement of disputes over fisheries?

As fisheries come into greater use, several kinds of conflicts can be anticipated. One, raised in Points 1 and 2 above, has to do with the definitions of species that might be accorded separate treatment. Another conflict, of particular difficulty during the transition period, will be that of the drawing of boundaries between neighboring and opposing coastal states. Persistent conflicts are likely to occur over the adoption of management measures and over the distribution of fisheries wealth.

As discussed in Chapter IV, some means for settling such disputes may be a vital element in the establishment of general fisheries regimes. The delegates at the UN Conference will want to consider whether the machinery should be adopted on a global or a regional basis; whether it should be compulsory or only advisory; and whether it should cover all fishery disputes or only certain kinds. Further discussion of these alternatives is contained in the aforementioned separate study in this series by Edward Miles.

7. Should the users of marine fisheries be required to contribute payments to an international fund? to a regional fund? If so, should this apply to fisheries beyond coastal state baselines? beyond territorial seas? or only beyond the limits of fisheries jurisdiction?

As noted in Chapter II, there are many advantages to the use of taxes, license fees, or other monetary means for the regulation of fisheries. These provide the most flexible means for control and they lead to the production and capture of economic rents that are generally dissipated by other regulatory measures. Furthermore, they do not necessarily mean a diminution in the quantity of fish produced or an increase in product prices. Thus, in terms of management and the optimum production of economic yields, payments for the use of fisheries are generally desirable.

However, if such payments are extracted, their distribution is a matter of negotiation rather than a matter of management. The distribution of revenues depends upon the degree to which it is desirable, or necessary, to accommodate the interests of the nonfishing states. (See Chapter III.)

SUMMARY

Most of the general principles discussed above have to do with the question of the degree of authority to be accorded to coastal states. This question is the fundamental fisheries issue that will face the delegates at the UN Conference. It also constitutes the major distinction between the proposals (as presently stated) for a species approach and those for a zonal approach to the resolution of fishery problems and the creation of new regimes. Under the former approach, coastal state authority is diffused and shared with foreign states. Under the latter, it is comprehensive in both scope and content.

One question that the delegates will have to answer is whether effective management of fisheries requires the exercise of a high degree of authority (leaving aside for the moment, where that authority should reside). Some may answer that the costs and difficulties of acquiring and exercising authority are very great, and that management can be effective even where authority is diffused. The general conclusion of this paper, however, is that a much higher degree of authority must be exercised than has been possible in the past, and that even though the costs of achieving it may be great, they are justified by the likely improvements in management.

If this conclusion is correct, it raises a second question: whether the authority can best be exercised by coastal states alone or by coastal states in cooperation with distant-water states. The answer to this depends in part upon the situation and the migratory characteristics of the different kinds of fish. In the case of the highly migratory oceanic species, it is clear that even the most extensive limits of jurisdiction claimed thus far provide the coastal states with insufficient controls over the stocks. This necessitates a wide sharing of authority. In other situations, however, the answer will depend upon the interests of the coastal states and their competence (either on their own or through the services of an international body) to manage the resources.

These questions have been raised in terms of the effectiveness of arrangements for management; that is, how fisheries can be managed so as to maximize the production of net benefits. Obviously, however, the delegates will view these questions in terms of their effects on the distribution of fisheries wealth as well as in terms of management. But it is to be hoped that their interests in receiving greater shares of wealth will not lead them to adopt general principles that will unduly diminish the net benefits that marine fisheries can produce.

Printed and bound by CPI Group (UK) Ltd, Croydon, CR0 4YY

22/10/2024

01777611-0003